MEDICAL DEVICES AND EQUIPMENT

ULTRASONOGRAPHY

TECHNOLOGY, DIAGNOSTIC APPLICATIONS AND POTENTIAL BENEFITS /RISKS

MEDICAL DEVICES AND EQUIPMENT

Additional books in this series can be found on Nova's website under the Series tab.

Additional e-books in this series can be found on Nova's website under the e-book tab.

MEDICAL PROCEDURES, TESTING AND TECHNOLOGY

Additional books in this series can be found on Nova's website under the Series tab.

Additional e-books in this series can be found on Nova's website under the e-book tab.

MEDICAL DEVICES AND EQUIPMENT

ULTRASONOGRAPHY

TECHNOLOGY, DIAGNOSTIC APPLICATIONS AND POTENTIAL BENEFITS / RISKS

ALEXANDER BURDENKO
EDITOR

NOVINKA

New York

For permission to use material from this book please contact us:
Telephone 631-231-7269; Fax 631-231-8175
Web Site: http://www.novapublishers.com

NOTICE TO THE READER

Additional color graphics may be available in the e-book version of this book.

Library of Congress Cataloging-in-Publication Data

Library of Congress Control Number: 2012954194
ISBN: 978-1-62417-536-7

Published by Nova Science Publishers, Inc. † New York

CONTENTS

PREFACE

This book explores the technology, diagnostic applications and potential benefits and risks of ultrasonography. Topics discussed in this compilation include the contemporary role of ultrasound in prostate biopsy; ultrasound in the management of ischemic stroke; the role of endoscopic ultrasonography in pancreatic solid masses; standard screening procedures of ultrasonographic examinations when surveying cervical lymph nodes in patients with oral squamous cell carcinoma; and pre-analytical error in ultrasonography.

Chapter 1 - Prostate biopsy is a crucial step in the diagnostic workup of prostate cancer. Two-dimensional grayscaletransrectal ultrasound (TRUS)-guided prostate biopsy has become the clinical standard for diagnosis of prostate cancer. This technique is most commonly used in a systematic fashion with a pre-determined biopsy scheme rather than specifically targeting suspicious lesions. The false negative rate of two-dimensional grayscale TRUS-guided biopsy is high, and therefore new lesion-targeted strategies have been developed. In attempt to overcome the limitations in both sensitivity and specificity of two-dimensional grayscale ultrasound, innovations such as colour and power Doppler ultrasound, contrast enhanced ultrasound, elastography, quantitative ultrasound, 3D ultrasound and MRI fused ultrasound have been implemented. In this review the authors will critically evaluate the current and prospective ultrasound guided prostate biopsy methodology.

Chapter 2 - Ultrasound is increasingly being used in the diagnosis, management and treatment of ischemic stroke. Linear array probes with frequencies of 5 to 10 MHz are used for extracranial studies, and sector or small hand-held 2 MHz probes for transcranial studies. Ultrasound is non-invasive, safe, painless, portable, quick, inexpensive, repeatable and accurate, but it requires adequate training; accreditation for physicians and laboratories are available. The extracranial arteries imaged include the common, internal and external carotid, vertebral, subclavian arteries bilaterally, and the brachiocephalic trunk. For transcranial studies, the ophthalmic artery, carotid siphon, terminal internal carotid, and proximal anterior, middle, posterior, basilar and intracranial vertebral arteries are evaluated. Brightness-(B)mode, intensity/power mode, colour-coded and Doppler mode are used; more recently, motion-(M)mode is also employed. Velocity parameters measured include the Peak Systolic, End-Diastolic and Mean velocities; resistivity parameters include the Gosling's Pulsatility Index and Pourcelot's Resistance Index. For safety, the Thermal Index cannot exceed 6, Mechanical Index 1.9, Spatial Peak Temporal Average Intensity 720 mW/cm2. Ultrasound is used to evaluate extracranial and intracranial arterial stenosis, collateral flow as a result of severe stenosis/ occlusion, pulsatile neck masses, subclavian steal, carotid dissection, Takayasu disease, right-to-left shunting, embolisation, intra-operative monitoring during carotid endarterectomy or CABG and post-surgery/stent assessment of stenosis, as well as cerebrovascular reserve to predict stroke risk. It is also used therapeutically to enhance thrombolysis (sonothrombolysis). Neurosonology provides an invaluable tool for the clinician in the management and prevention of ischemic stroke.

Chapter 3 - Endoscopic ultrasonography is a combination of endoscopic view and ultrasonography. It was developed to improve diagnostic accuracy by directly imaging the target lesion via the gastrointestinal lumen. The examination of the pancreatic gland is one of the most common indications of the endoscopic ultrasonography because of the vicinity of the endoscopic ultrasonography probe located in the stomach or duodenum to the pancreas. In fact, endoscopic ultrasonography has been considered the most accurate imaging technique in terms of detection rate and locoregional staging, particularly in tumors smaller than 2 cm in size. This technique may also be useful for clarifying abnormalities detected by other techniques such as the computed tomography or the magnetic resonance imaging. Diagnosis of solid lesions located in the pancreatic gland are usually challenging for the clinician. Although most of them are adenocarcinomas, diagnosis is not always straightforward and a cytological or histological sample is sometimes mandatory because clinical management may be different depending on the diagnosis. Unusual solid lesions such as pancreatic metastasis, pancreatic lymphoma or neuroendocrine tumors account for 10% of the pancreatic masses and should be suspected based on patient´s age, family history of pancreatic cancer, past personal history of cancer and location of the tumor.

Other conditions such as focal chronic pancreatitis or autoinmune pancreatitis may mimic a pancreatic adenocarcinoma and in these cases fine needle aspiration biopsy is required to rule out cancer. A cytological diagnosis is mandatory as a proof of malignancy in non-resectable pancreatic tumors before initiation of chemotherapy. In this way, endoscopic ultrasonography has advantages over other imaging techniques when the authors are dealing with solid pancreatic lesions including real time imaging and a lower risk of complications when an endoscopic ultrasound -guided fine needle aspiration cytology is required. In this chapter the authors review the indications of endoscopic ultrasonography and endoscopic ultrasound -guided fine needle aspiration cytology in solid pancreatic masses including diagnosis, indications of cytology and its usefulness in staging.

Chapter 4 - In the authors' previous study, the authors elucidated that scans that included the thyroid gland should be performed when surveying the cervical lymph nodes for metastasis during ultrasonographic (US) examinations in patients with oral squamous cell carcinoma (SCC). However, no standard procedures have yet been established for US examinations to survey the cervical lymph nodes in patients with oral SCC. Therefore, the authors developed a standard US procedure and evaluated the legitimacy and drawbacks associated with it. For the authors' US scan procedure to survey cervical lymph nodes in patients with oral SCC, five axial, two cross-sectional, and one coronal tomographic scan using a probe with a major axis of 10 cm and a minor axis of 2 cm were used for the lateral necks of Japanese patients. If the resulting images were thought to be insufficient by the examiner, then some additional and repeated US examinations were performed.

In the present study, the authors analyzed the detection rates of cervical lymph nodes with a major axis over 8 mm using the authors' US procedure for patients with oral SCC as a gold standard, using both magnetic resonance (MR) images and computed tomography (CT) images, which were obtained by both an expert and by a less-experienced examiner to examine how difficult the procedure was to follow and how accurate the results were. At the same time, the mean scanning hours spent by the expert and less-experienced examiner per patient was assessed when using the authors' US procedure. In addition, the ink-smeared areas on the volunteer's neck as a result of the authors' US procedure were analyzed. The detection rates of enlarged cervical lymph nodes using the authors' US procedure were almost 95% for both the expert and the less-experienced examiner. In addition, the overlapped area of ink smeared on a standard volunteer's neck using the authors' standard US procedure perfectly matched the lymph nodes in the cervical areas and thyroid glands. The time spent scanning the patients by the expert ranges from fifteen to twenty minutes, but the less-experienced examiner spent from twenty to twenty-five minutes per patient. Therefore, the authors' US procedure for surveying the cervical lymph nodes, including thyroid glands, in patients with oral SCC can be recommended as a relatively good method, but inexperienced examiners should be prepared to spend additional time analyzing the scans.

Chapter 5 - Ultrasonography is widely used as diagnostic technology in medicine at present. It is no doubt that ultrasonography has a lot of benefits, however, the risk in using this technique has to be mentioned. Based on the laboratory medicine principle, the concern on the pre-analytical phase for using diagnostic ultrasonography should be mentioned. The error in pre-analytical phase can be expected and this can be risk. Of interest, pre-analytical error in ultrasonography is usually not mentioned and it is an actual forgotten issue. This short commentary will summarize and discuss on this specific topic.

In: Ultrasonography
Editor: Alexander Burdenko

ISBN: 978-1-62417-536-7

Chapter 1

THE CONTEMPORARY ROLE OF ULTRASOUND IN PROSTATE BIOPSY: A CRITICAL EVALUATION OF CONVENTIONAL METHODS AND INNOVATIONS

Alyssa Louis and Uri Lindner*
Division of Urology, Department of Surgical Oncology,
University Health Network, Toronto, Canada

ABSTRACT

Prostate biopsy is a crucial step in the diagnostic workup of prostate cancer. Two-dimensional grayscale transrectal ultrasound (TRUS)-guided prostate biopsy has become the clinical standard for diagnosis of prostate cancer. This technique is most commonly used in a systematic fashion with a pre-determined biopsy scheme rather than specifically targeting suspicious lesions. The false negative rate of two-dimensional grayscale TRUS-guided biopsy is high, and therefore new lesion-targeted strategies have been developed. In attempt to overcome the limitations in both sensitivity and specificity of two-dimensional grayscale ultrasound,

* Corresponding Author: Uri Lindner, University Health Network, Princess Margaret Hospital, Suite 3 130, 610 University Avenue, Toronto, Ontario, Canada, M5G 2M9, Telephone: +1 647 774 1578, Fax: +1 416 340 4275, E-mail: lindneruri@gmail.com.

innovations such as colour and power Doppler ultrasound, contrast enhanced ultrasound, elastography, quantitative ultrasound, 3D ultrasound and MRI fused ultrasound have been implemented. In this review we will critically evaluate the current and prospective ultrasound guided prostate biopsy methodology.

BACKGROUND

Prostate Cancer remains the most commonly diagnosed cancer in men globally and the 6th leading cause of cancer death [1]. While prostate cancer is a highly curable disease with excellent 5-survival rates exceeding 93%, appropriate treatment decision relies heavily on accurate staging [2]. Depending on the estimated risk of cancer progression, men may be candidates for active surveillance, focal therapy or radical, whole-gland therapy. Radical therapies, including external beam radiotherapy, brachytherapy and radical prostatectomy, aim to remove or destroy the entire prostate gland. However, recent reports suggest that the number of men that need to be treated to prevent a single prostate cancer death is high, with a range of 13-42 individuals [3]. This suggests a high degree of overtreatment, at substantial morbidity and expense to the patient. In order to overcome this issue, active surveillance protocols were developed in which men defer treatment until clinical evidence of progression is detected, thus prolonging, sometimes indefinitely, the time without the morbidity of treatment. Active surveillance includes close observation of prostate specific antigen (PSA), digital rectal exam (DRE) and serial prostate biopsy. Another more recent strategy called focal therapy aims to immediately treat only clinically significant prostate tumours while leaving the rest of the gland intact, thus providing oncologic control without the morbidity of radical treatment [4]. While the conservative strategies of active surveillance and focal therapy could theoretically solve the over-treatment paradigm, they rely heavily on accurate prostate cancer staging.

Contemporary methods of diagnosis and staging are limited. The widely used PSA test is nonspecific to prostate cancer. The threshold for "normal" PSA is generally around 4ng/ml, yet men without prostate cancer may have elevated PSA due to benign enlargement of the prostate or prostatitis. In addition, approximately 25-35% of men with PSA <4ng/ml at diagnosis harbour aggressive prostate cancer [5,6]. A novel biomarker PCA3 shows promise, but its use in prostate cancer screening and diagnosis has not been fully established [7]. Digital rectal examination has both low specificity and sensitivity [8].

While maintained as the gold standard of prostate cancer diagnosis, the optimal biopsy regimen is still not agreed upon. Prostate biopsy is most often performed under ultrasound guidance and biopsies are obtained from the different "zones" of the prostate as established by convention. Sextant biopsy schemes obtain one or more cores from the right and left side of the upper, middle and lower zones. This blind sampling has a high false negative rates, with approximately 60% sensitivity [9].

Further, it is well established that prostate tumours scored as Gleason 6 (3+3) on biopsy were upgraded based on specimen findings at a rate as high as 62% [10]. The incorrect classification of tumours as Gleason 6 has serious implications for active surveillance and focal therapy protocols which often exclude individuals with clinically significant Gleason 7 disease. In attempt to improve the discordance rates between biopsy and specimen Gleason scores and provide more accurate grading, new lesion-targeted biopsy protocols have been developed. Here we discuss the limitations of contemporary methods of two-dimensional grayscale ultrasound guided biopsy and recent innovations.

Two Dimensional Transrectal Ultrasound and Extended Biopsy Schemes

Two dimensional grayscale transrectal ultrasound (TRUS) guided biopsy is the standard biopsy procedure in the majority of centres utilizing a pre-defined biopsy scheme to sample the prostate methodically but without targeting suspicious areas. The TRUS-guided sextant biopsy gained popularity over digitally guided biopsy in 1989 when Hodge and colleagues demonstrated improved detection [11]. Sextant biopsy schemes obtain six samples from the prostate, obtaining the sample cores from the right and left base, middle and apical regions of the prostate. The six cores were originally taken from the mid-parasagittal plane. However, in order to obtain better sampling of the peripheral zone, which is where approximately 70% of tumours arise, Stamey recommended more lateral sampling at the expense of transitional zone sampling [12]. Presti and colleagues reported that an 8-core biopsy scheme that sampled both the transitional zone and the peripheral zone of the middle region of the prostate gland reduced the percentage of missed tumours from 20% to 11% compared to centrally directed sextant biopsy [13]. Further,

Kawata and colleagues found that laterally directed biopsy in a "modified fan shape" increased detection of clinically threatening tumours to 91% from 66% with the standard sextant biopsy [14].

A number of groups investigated immediate or delayed re-biopsy of men with a standard sextant protocol. They reported that approximately 20-30% of the tumours they detected were found only on the second biopsy [15,16]. Further, sextant biopsy protocols tend to miss a greater proportion of tumours in men with large prostate glands or with small volume disease as discussed by Presti [17]. Karakiewicz and colleagues reported significant differences in detection rates based on gland size, ranging from 40% in men with prostates <20cc to 10% in men with prostates 80-89.9cc [18]. Chen and colleagues reported that men with prostates >50g had a higher proportion of small volume <0.5cc disease, further evidence for development of extended biopsy schemes [19]. A mathematical model conducted by Stricker and colleagues also demonstrated that increasing the number of biopsies would improve detection of fixed volume percent cancers [20].

Eskew and colleagues introduced a 5-region extended biopsy scheme involving 13 cores placed evenly from the midline outwards. They reported the greatest improvement in detection for men with PSA<10, while 90% of tumours in men with PSA>10 were detectable by sextant biopsy alone [21].

Scattoni and colleagues proposed that the number and location of cores should differ based on digital rectal exam results, prostate volume and patient age. With the results from a systematic 24-core protocol they determined that in order to detect 95% of tumours, the maximum number of cores necessary was 16 for patients with a negative DRE, small prostate volume and age less than 65 and 10 cores for men with a positive DRE [22]. These results suggest a minimum of 8 peripheral zone samples, with variable positions and number of samples of the apex and transitional zone depending on individual clinical characteristics.

Saturation biopsies may result in over 30% increase in cancer detection rate in men with a previous negative biopsy [23,24]. Onik and colleagues described a protocol in which cores were taken every 5mm within the prostate, and reported a false negative in only one of 36 patients biopsied [25]. Barzell and colleagues performed mapping biopsy to identify focal therapy candidates with a mean of approximately 66 cores and reported 100% sensitivity [26]. However, this necessitates general anesthesia and are costly.

Two dimensional grayscale imaging is most useful for sampling anatomic locations rather than specifically targeting lesions as approximately 40% of prostate tumours are isoechoic [27]. The false negative rate when compared to a mapping biopsy was reported as high as 47% in the re-biopsy setting [26].

DOPPLER ULTRASOUND

Doppler ultrasound utilizes the Doppler effect in order to estimate velocity of blood flow. The frequency shift of the reflected waves provides enough information to determine the speed and direction of flow through a blood vessel. Due to the hypervascularity of the tumour microenvironment this technique is able to resolve lesions from normal prostate tissue. Doppler may be performed with or without the microbubble contrast agents described separately.

Power Doppler, a technique that displays the total energy of the signal by integrating the Doppler signal, was able to yield a 35% detection rate on first biopsy as reported by Remzi and colleagues. However, prostate cancer was detected in 10% of normal appearing cases. Sensitivity and specificity were 83% and 79%, with positive and negative predictive values of 88% and 90% respectively [28]. Kimura and colleagues reported on a similar protocol, with overall detection rate of 47%, with sensitivity and specificity of 90% and 70% [29]. However, when Yang and colleagues compared power Doppler with contrast enhanced ultrasound and grayscale ultrasound they reported that contrast enhanced ultrasound was superior in terms of sensitivity and specificity and could have saved a significant number of patients from undergoing unnecessary biopsy [30].

Taverna and colleagues randomized men to standard 2D grayscale TRUS, colour Doppler or colour Doppler with Sonovue in 300 patients. Prostate cancer detection rate was approximately 29% overall, with no significant differences in detection between the three groups, with low sensitivity and specificity for the different protocols [31]. Colleselli and colleagues reported no significant detection differences overall when comparing contrast enhanced colour Doppler with grayscale systematic biopsy [32]. In a comparison of sextant plus laterally directed biopsies with sextant plus colour Doppler targeted biopsies, Kravchick and colleagues reported no significant differences [33].

Mitterberger and colleagues performed a randomized trial comparing contrast enhanced colour Doppler ultrasound 5 core targeted biopsy with 10-core systematic grayscale biopsy. They reported significantly improved detection rate of 32% in the targeted biopsy protocols compared to 26% in the systematic group and reported higher detection efficiency reflected in the percentage of overall cores containing cancer with the targeted approach (16% compared to 7%) [34]. This confirmed the earlier work of Frauscher and colleagues, which demonstrated significantly improved detection rate with targeted biopsy [35]. Pelzer and colleagues reported that SonoVue enhanced colour Doppler targeted biopsy was approximately 3 times more likely to detect cancer compared to systematic biopsy, however the highest detection rate was a combined systematic and targeted approach [36].

Mitterberger and colleagues later reported improved detection when Dutasteride, a 5α-reductase inhibitor, was administered in a 14-day course prior to contrast enhanced colour doppler biopsy protocol, however this was compared to systematic grayscale biopsy and not directly compared with the same protocol without Dutasteride. The authors confirmed their hypothesis that by reducing blood flow to benign prostate tissue, which is responsive to Dutasteride, the bloodflow to the tumour tissue would remain high and offer better contrast [37]. This protocol offers the advantage of distinguishing between benign and malignant lesions for targeted biopsy, but further randomized controlled trial is warranted.

Another interesting pharmacologic intervention to augment power Doppler ultrasound was evaluated by Morelli and colleagues who administered vardenafil, a phosphodiesterase-5 inhibitor in order to increase blood flow to the tumour microvasculature. They reported significant higher detection with the vardenafil-augmented power Doppler ultrasound compared to systematic and contrast enhanced power Doppler [38].

ELASTOGRAPHY

Prostate tumours are thought to possess different mechano-elastic properties than normal prostate parenchyma. Strain imaging, or elastography was described by Ophir and colleagues as a method of differentiating the elastic moduli of tissues following external application of stress [39]. As reported by Krouskop and colleagues, there is a significant difference in stiffness between normal and neoplastic prostate tissue [40]. A study performed by Sumura and colleagues comparing pre-operative elastography studies with radical prostatectomy specimens reported a 73% detection rate for tumours <1mL, and 100% detection rate for tumours >5mL [41].

Cochlin and colleagues performed elastography in 100 patients by manually "flicking" the ultrasound transducer against the anterior wall of the rectum to produce tissue deformation. The results were used to identify abnormal regions to direct targeted cores in addition to the sextant protocol. The authors reported that the majority of cancers detected by elastography were also detected by grayscale imaging and therefore elastography did not significantly add to the diagnostic accuracy or staging accuracy of the biopsy protocol, however three additional diagnoses were made that would have been missed with sextant biopsy alone [42].

Rather than targeting suspicious areas demonstrated by elastography, Konig and colleagues attempted to predict which cores of a conventional sextant biopsy would be positive based on the elastogram being displayed simultaneously during the procedure. They reported correct prediction in only 51% of cancerous specimens. They suggested that benign alternations to the prostate and prostatitis led to problematic interpretation of the elastograph, and advised that conventional grayscale ultrasound be used alongside elsastography in all cases [43].

Pallwein and colleagues compared elastography-guided targeted biopsy with ten core systematic biopsy. Cancer detection by targeted biopsy was approximately 30%, and 10% of the overall cohort was diagnosed with targeted biopsy whose cancers would have not been detected with the systematic protocol [44].

Brock and colleagues randomized men to grayscale and elastography guided systematic biopsy in which lesions in each target sector were targeted if detected. They reported higher sensitivity of elastography for detecting prostate cancer compared to grayscale ultrasound, however this was not enough to eliminate the need for systematic biopsy [45].

Some modification and optimization of current elastography technique is warranted. Among suggestions by Pallwein and colleagues were adapting a balloon system to the probe to indicate the force of compression by the prostate [46].

Contrast Enhanced Grayscale Ultrasound

Contrast Enhanced Ultrasound (CEUS) is 2D greyscale ultrasound with intravenous contrast agent. The contrast agents that are being used commercially are: Sonovue (Bracco, Milan, Italy), an aqueous suspension of sulphur hexafluoride microbubbles surrounded by a phosopholipid monolayer shell, Levovist (Shering Berlin, Germany), a galactose-palmitic acid, Definity (Bristol-Myers Squibb Medical Imaging, Billerica, USA) perflutren lipid microspheres, EchoGen (SONUS Pharmaceuticals, Bothell, USA) perfluorocarbon microbubbles and Imagent (Imcor, San Diego, USA) lipid microspheres. These contrast agents are prone to destruction when exposed to low frequency and high power ultrasound and some trade-off for signal optimization in necessary. The backscattering caused by the contrast agent occurs in harmonics or subharmonics of the frequency sent by the probe and therefore the signal to noise ratio can be dramatically increased when these return frequencies are selected [47].

The high density and random distribution of capillaries in prostate carcinoma causes a faster and higher intensity enhancement than normal prostate parenchyma [48]. Tang and colleagues reported enhancement of approximately 95% of malignant lesions and 70% of benign lesions, with a sensitivity of 71% and specificity of 65%. CEUS alone is not able to distinguish benign from malignant lesions, CEUS can indicate specific areas to target for biopsy [49].

Halpern and colleagues reported on a series of 40 patients indicated for biopsy who underwent Definity enhanced US for systematic and lesion-directed biopsy, yielding a 40% detection rate. Sensitivity of imaging was approximately 40% and specificity was approximately 80%. The predictive value of an enhancing focus for a positive biopsy was 50%. The authors reported a higher detection rate with two cores per suspicious site [50]. With a similar Sonovue enhanced US protocol, Yang and colleagues reported a detection rate of 55% in 115 patients referred to biopsy for elevated PSA. They performed lesion directed and sextant biopsy noting enhancing sites in 43% of patients overall.

Linden and colleagues reported on a series of 60 patients evaluated with MicroFlow Imaging of Definity contrast directed biopsy. This is a flash replenishment technique in which a high powered ultrasound pulse is used to destroy microbubbles and then their replenishment is measured, improving the ability to visualize microbubbles in the tumour capillaries. Abnormal enhancement was seen in 92% of patients and the detection rate for prostate cancer was 30% overall, with the directed cores twice as likely to detect cancer than the systematic cores, however only 3% of patients were diagnosed on directed biopsy alone [51]. A similar protocol was evaluated by Aigner and colleagues with Sonovue flash replenishment followed by directed and systematic biopsy. They reported a 100% sensitivity for flash replenishment to detect cancer, however there was also a false positive rate of approximately 20% [52].

Computer Assisted Quantification Methods

A number of quantification systems have been developed in which computerized analysis of raw ultrasound data identifies locations where there is a high probability for cancer, and therefore identifies areas in which to direct biopsy cores. Current systems include Histoscanning, Automated Urologic Diagnostic Expert (AUDEX) and computerized transrectal ultrasound (CTRUS). These systems analyze data from two dimensional grayscale ultrasound. Histoscanning is a technology that consists of characterization algorithms applied on backscattered 3D ultrasound data. Brackman and colleagues developed the system based on radical prostatectomy specimens and reported results from a pilot study. They found significant concordance between Histoscanning estimated tumour location and diameter as measured in the pathological specimen. Later Simmons and colleagues reported sensitivity of approximately 90% for cancer foci of ≥0.2mL in patients with biopsy proven cancer [53]. The AUDEX system was developed by finding correlation between 2D grayscale ultrasound features and histopathologic results from biopsy cores [54]. When the system was initially evaluated in radical prostatectomy specimens, the sensitivity and specificity was 75% and 78% as reported by Giesen and colleagues [55]. However, Beerlage and colleagues reported on a larger group of 62 patients and reported sensitivity of 85%, specificity of 18% and diagnostic accuracy of 57% rendering the AUDEX system inappropriate for stand-alone clinical use [56]. However, this still may be of use for directing biopsy. CTRUS targeted biopsy was evaluated in a series of patients who had multiple negative systematic biopsies. Loch reported 50% detection rate at the targeted re-biopsy, which used landmarks such as calcifications and cysts to help direct the biopsy needle. Recently, the CTRUS protocol was adapted so that image analysis could be performed remotely offsite, allowing multicentre participation [57]. A current limitation to the computerized protocols is that the biopsy target is not identified in real time, rather the data must first be processed with the CTRUS software. This necessitates a minimum of two ultrasound visits to obtain a biopsy.

Three Dimensional Ultrasound Guided Biopsy

Three dimensional (3D) ultrasound consists of acoustic waves that are directed at multiple angles through either mechanical or manual sweeping, or

2D transducer arrays that can intrinsically capture 3D data [58]. The reflections may be reconstructed in real-time to create the 3D image, referred to as 4D ultrasound imaging. This technique was investigated for prostate imaging by Elliot and colleagues, and Tong and colleagues [59,60]. Recently a number of groups have investigated the feasibility of 3D ultrasound guidance for prostate biopsies to facilitate superior anatomic context compared to 2D ultrasound. The third dimension is beneficial for volume calculation, target planning, navigation and motion compensation [61].

A number of commercially available systems are under evaluation for prostate biopsy usage, including the Voluson (GE Healthcare, Chalfont St Giles, England), Artemis (Eigen, Grass Valley, USA), TargetScan (Envisioneering Medical, St Louis, USA) and The Duke/VMI 3D scanner (Volumetrics Medical Imaging, Durham, USA). The Voluson 5-8MHz features 360° radial scanners with an end viewing transducer for imaging in the axial, sagittal and semi-coronal planes. The Artemis system features an end firing probe mounted on an arm with four degrees of freedom, which pivots 160° about while positioned in order to scan the entire prostate. The system provides real-time locations of the probe so that the biopsy needle's location with respect to the prostate is known. The TargetScan uses anatomic landmarks of the prostate to generate computer-directed biopsy targets that are obtained by a physician-inserted needle. The system calculates and displays the accuracy of the needle placement with respect to the target.

The Volusonendorectal probe was also tested in the context of prostate biopsy with phantoms by Long and colleagues. Compared to standard 2D ultrasound, the 4D protocol yielded significant improvements on accuracy and reproducibility of the ideal target zone [62].

Abul and colleagues evaluated the Voluson adapted with an automatic biopsy core device on a series of 64 patients and reported that 80% had sonographic abnormality and an overall 23% detection rate with sextant biopsy and additional cores for abnormal echogenic regions if present. They reported that while hypoechogenicity was significantly associated with malignancy, 27% of men with positive biopsy were isoechoic [63].

Cool and colleagues compared 2D ultrasound, 3D ultrasound and 3D ultrasound guided towards specific in a re-biopsy simulation. They reported a 100% successful biopsy rate of clinically significant tumours >0.5cm when the target was actively displayed to the user intra-procedure along with visual feedback of the needle path. 3D guided ultrasound reduced biopsy error compared to conventional 3D and 2D ultrasound [64]. It is worth noting that this targeting success applies only to the context of re-biopsy after previous histopathologic detection of an abnormality.

In order to overcome prostate deformation and therefore target displacement by the probe operator, Bax and colleagues developed a 3D navigation system for prostate biopsy with a passive arm to minimize motion following probe positioning. They reported mean biopsy core error in phantom studies of 1.8mm [65]. This is acceptable given that this would still be within the diameter of a 0.5cm clinically significant mass.

An interesting innovation developed by Narayanan and colleagues involves fast registration of the 3D-TRUS image with a 3D prostate cancer atlas which displays colorimetric regions with high statistical likelihood for cancer detection. This atlas was generated from the tumour locations within 138 radical prostatectomy specimens. Their fast registration in approximately 1 second yielded intersection to a union ratio for surface overlap >0.85. The cancer detection rate was 85-90% with 7-12 cores compared to the 70% detected with conventional sextant biopsy [66].

Fei and colleagues developed a molecular image-directed 3D ultrasound guided biopsy system. They aim to develop a method to target areas identified as suspicious on PET/CT imaging with the ^{18}F-FACBC tracer, which is a promising prostate cancer marker [67]. While their system remains subclinical, they were able to perform image fusion with suitable overlap, and further investigation is warranted [68].

The 3D ultrasound alone may improve upon the spatial constraints of 2D TRUS and improve targeting to target sites within the prostate, however this offers little improvement in the specificity of ultrasound for malignant tumours.

Optoacoustic Imaging

The photoacoustic effect is a phenomenon in which light absorption leads to rise in temperature, material expansion and generation of an acoustic wave. Optoacoustic imaging captures this phenomenon by delivering a short laser pulse and detecting the incident sound wave using a modified ultrasound probe. Compared to benign prostate tissue, prostate tumours have more extensive blood vessel network that causes these regions to have higher haemoglobin levels. Hemoglobin strongly absorbs light in the near-infrared range and therefore the blood rich malignant tissues have enhanced absorption for optoacoustic imaging [69].

Yaseen and colleagues evaluated optoacoustic imaging in vivo in a canine prostate. They simulated tumour foci by creating bloody lesions within the prostate, which were visible with optoacoustic imaging but not on grayscale ultrasound. Other structures that were visible included the urethra and prostate capsule, which may serve as landmarks. However, aside from these trials occurring in excised prostates, their haemorrhagic lesion likely contained a higher concentration of haemoglobin than expected in a tumour [70]. Wang and colleagues also performed optoacoustic imaging of bloody lesions induced by injecting fresh canine blood directly into the prostate parenchyma of anesthetised dogs. The optoacoustic imaging was able to detect the lesions whereas grayscale ultrasound was not [71].

Optoacoustic imaging of the prostate remains subclinical and there are no commercially available endorectal probes containing a laser capable of delivering the near-infrared light. Also, the sensitivity and specificity for malignant prostate tumours has not yet been determined. However, this technique holds promise.

MRI-Fused Ultrasound

MRI has been used extensively to evaluate prostatic diseases and is renowned for its detailed soft tissue contrast and spatial resolution. Prostate cancer localization began with the usage of T2 weighted imaging. On T2 the prostate gland is clearly localized and prostate tumours display low signal strength. However, other conditions such as prostatitis, haemorrhage and scarring have similar appearance. Sensitivity and specificity range from 54-85% and 46-96% respectively [72-74]. MRI may offer additional information about the prostate tissue through diffusion weighted imaging which measures the diffusion of water in tissues. Prostate tumours are characterized by high cell density which restricts diffusion of water leading to low signal intensity on apparent diffusion coefficient maps [75]. Dynamic contrast-enhanced MRI for prostate studies usually involves intravenous injection of a contrast agent, usually Gadoliunium. Early enhancement and washout is characteristic of malignant prostate tumours, and it has recently been demonstrated that washout gradient is associated with tumour aggressiveness [76]. Another method, magnetic resonance spectroscopic imaging measures tissue metabolite levels including choline, citrate creatine and various polyamines. Prostate cancer typically features high concentration of choline and low concentration of citrate and other polyamines compared to benign tissue. The ratios of citrate, choline and creatine have been correlated with cancer aggressiveness [77].

MRI guided prostate biopsy was reported in 2000 by D'Amico and colleagues [78]. Since then a number of other groups have designed protocols for MRI-guided lesion-targeted biopsy, however several drawbacks exist. First, procedure times for MRI are higher than those of a typical ultrasound guided procedure and MRI time may be quite costly. Also, specialized MRI compatible biopsy needles and other equipment may be necessary. Finally the process of obtaining a biopsy with the patient in the MRI may be challenging due to the ergonomics of a closed-bore magnet and may necessitate development of manual assistive devices or even robotics. To overcome such challenges, MRI fused ultrasound was developed. A pre-biopsy MRI is obtained and then registered to ultrasound images in order to guide biopsy using the suitable anatomic information provided by grayscale ultrasound with the superior lesion localization offered by MRI.

Ewertson and colleagues performed a trial with prostate phantoms and found that a 1cm target sphere visible on previously obtained MRI but not ultrasound could be biopsied successfully on the first try in 73% of cases in a mean 2 minutes spent on each biopsy target following registration. They noted however that including the initial co-registration, the total procedure time was longer than a sextant biopsy protocol [79]. Xu and colleagues also performed phantom studies and reported a clinically acceptable spatial accuracy of approximately 2.4mm [80]. Another phantom study performed by Kuru and colleagues reported a targeting error of 0.83mm [81].

Singh and colleagues reported feasibility of MRI fused ultrasound guided prostate biopsy during fiducial placement for radiotherapy [82]. Kaplan and colleagues similarly described feasibility of real time MRI-ultrasound guided stereotactic prostate biopsy to detect local recurrence following external beam radiotherapy [83]. In order to facilitate registration of the MRI and ultrasound prostate images, the deformation caused by the TRUS probe must be overcome. Ukimura and colleagues developed a plane biplane rigid registration technique with a plastic probe developed to resemble the TRUS probe placed in the rectum during pre-biopsy MRI [84].

Hadaschik and colleagues reported on 106 men referred for biopsy on suspicion for prostate cancer. They used the BiopSee (Medcom, Damstadt, Germany) stereotactic prostate biopsy system which necessitated 3T multiparametric MRI. They used a transperineal approach to overcome deformation caused by the TRUS probe and the imprecision of manually placing a TRUS probe. This necessitated the use of anesthesia. In MRI suspicious lesions, tumour detection rate was 96% and the overall detection rate was approximately 60% of men enrolled in the study with combined lesion-targeted and systematic biopsy [85].

Pinto and colleagues reported that 3T multiparametric MRI fused 3D ultrasound lesion targeted biopsy in 101 patients detected significantly more cancer per core than standard 12-core systematic biopsy (21% and 12% respectively). They also demonstrated detection rates differed for men for men based on the number of suspicious MRI sequences [86].

Rud and colleagues implemented a protocol with the Urostation (Koelis, Grenoble, France) in which 90 patients were biopsied with TRUS and T2 weighted MRI soft image fusion. They reported positive biopsy in 52% of the cancer suspicious lesions, with a maximum of 91% for high suspicion targets [87]. Another group used the Artemis (Eigen, Grass Valley, USA) to produce MRI fused 3D ultrasound to guide lesion targeted biopsy. They reported significantly higher detection rate with lesion targeting compared to systematic biopsy [88].

A different application of MRI fused ultrasound was reported by Turkbey and colleagues, who tracked the location of TRUS guided sextant biopsy cores. Fifty patients underwent pre-biopsy T2 weighted MRI, which was fused to 3D ultrasound obtained during the biopsy [89]. Knowledge of the anatomic location of each biopsy core may be of value for men who require re-biopsy for diagnostic purposes, surveillance or lesion-targeted focal therapies.

MRI fused ultrasound offers advantages over either MRI directed, and ultrasound directed biopsy alone, however the need for a pre-biopsy MRI may not always be feasible due to cost, waiting times and contra-indications. Baseline MRI may have a role in men with negative systematic biopsy with persistent suspicion of cancer or men on surveillance.

Caveats and Cautions

The current trend in prostate biopsy is toward lesion identification and targeting for biopsy [90]. Prostate cancer, however, remains a multifocal disease and smaller foci may be easily missed given the imperfect spatial resolution and sensitivity of current imaging modalities. Targeting may miss such foci, which may be favourable given the growing tendency towards active surveillance of insignificant disease, however without biopsy confirmation that small lesions harbour low grade disease, there can be no certainty that they are truly indolent. Also, the review we presented confirms that some lesions are missed even with advanced imaging techniques. Until these issues are resolved, or better biomarkers are developed and implemented, systematic biopsy will remain a valuable diagnostic tool.

A large number of studies we reviewed used sophisticated technologies to perform biopsy, the cost of which cannot be overlooked. Also, there will eventually be need for trained individuals outside of the research teams to perform the newly developed biopsy protocols, and the learning curve should not be underestimated. These issues may remain a caveat for widespread implementation, and the results reported in the literature may remain confined to those obtained in academic centres of excellence for the time being.

While randomized controlled trial remains the best method for evaluating new protocols, most are compared with systematic sextant biopsy. There is need for comparison between the new protocols both within and among academic centres, as there is considerable variability in the reported detection rates, sensitivity and specificity of each new modality.

REFERENCES

[1] Jemal A, Bray F, Center MM, Ferlay J, Ward E, Forman D. Global cancer statistics. *CA Cancer J Clin.* Mar-Apr 2011;61(2):69-90.

[2] Welch HG, Schwartz LM, Woloshin S. Are increasing 5-year survival rates evidence of success against cancer? *Jama.* Jun 14 2000;283(22):2975-2978.

[3] Abdollah F, Sun M, Schmitges J, et al. Survival benefit of radical prostatectomy in patients with localized prostate cancer: estimations of the number needed to treat according to tumor and patient characteristics. *J. Urol.* Jul 2012;188(1):73-83.

[4] Ahmed HU, Hindley RG, Dickinson L, et al. Focal therapy for localised unifocal and multifocal prostate cancer: a prospective development study. *The lancet oncology.* Jun 2012;13(6):622-632.

[5] Bonet M, Merglen A, Fioretta G, et al. Characteristics and outcome of prostate cancer with PSA <4 ng/ml at diagnosis: a population-based study. *Clin. Transl. Oncol.* May 2009;11(5):312-317.

[6] Thompson IM, Pauler DK, Goodman PJ, et al. Prevalence of prostate cancer among men with a prostate-specific antigen level < or =4.0 ng per milliliter. *The New England journal of medicine.* May 27 2004;350(22):2239-2246.

[7] Goode RR, Marshall SJ, Duff M, Chevli E, Chevli KK. Use of PCA3 in detecting prostate cancer in initial and repeat prostate biopsy patients. *The Prostate.* May 14 2012.

[8] Schroder FH, van der Maas P, Beemsterboer P, et al. Evaluation of the digital rectal examination as a screening test for prostate cancer. Rotterdam section of the European Randomized Study of Screening for Prostate Cancer. *Journal of the National Cancer Institute.* Dec 2 1998;90(23):1817-1823.

[9] Terris MK. Sensitivity and specificity of sextant biopsies in the detection of prostate cancer: preliminary report. *Urology.* Sep 1999;54(3):486-489.

[10] Pinthus JH, Witkos M, Fleshner NE, et al. Prostate cancers scored as Gleason 6 on prostate biopsy are frequently Gleason 7 tumors at radical prostatectomy: implication on outcome. *J. Urol.* Sep 2006;176(3):979-984; discussion 984.

[11] Hodge KK, McNeal JE, Terris MK, Stamey TA. Random systematic versus directed ultrasound guided transrectal core biopsies of the prostate. *J. Urol.* Jul 1989;142(1):71-74; discussion 74-75.

[12] Stamey TA. Making the most out of six systematic sextant biopsies. *Urology.* Jan 1995;45(1):2-12.

[13] Presti JC, Jr., Chang JJ, Bhargava V, Shinohara K. The optimal systematic prostate biopsy scheme should include 8 rather than 6 biopsies: results of a prospective clinical trial. *J. Urol.* Jan 2000;163(1):163-166; discussion 166-167.

[14] Kawata N, Miller GJ, Crawford ED, et al. Laterally directed biopsies detect more clinically threatening prostate cancer: computer simulated results. *Prostate.* Oct 1 2003;57(2):118-128.

[15] Keetch DW, Catalona WJ, Smith DS. Serial prostatic biopsies in men with persistently elevated serum prostate specific antigen values. *J. Urol.* Jun 1994;151(6):1571-1574.

[16] Levine MA, Ittman M, Melamed J, Lepor H. Two consecutive sets of transrectal ultrasound guided sextant biopsies of the prostate for the detection of prostate cancer. *J. Urol.* Feb 1998;159(2):471-475; discussion 475-476.

[17] Presti JC. Prostate biopsy: current status and limitations. *Rev. Urol.* Summer 2007;9(3):93-98.

[18] Karakiewicz PI, Bazinet M, Aprikian AG, et al. Outcome of sextant biopsy according to gland volume. *Urology.* Jan 1997;49(1):55-59.

[19] Chen ME, Troncoso P, Johnston D, Tang K, Babaian RJ. Prostate cancer detection: relationship to prostate size. *Urology.* Apr 1999;53(4):764-768.

[20] Stricker HJ, Ruddock LJ, Wan J, Belville WD. Detection of non-palpable prostate cancer. A mathematical and laboratory model. *Br. J. Urol.* Jan 1993;71(1):43-46.

[21] Eskew LA, Bare RL, McCullough DL. Systematic 5 region prostate biopsy is superior to sextant method for diagnosing carcinoma of the prostate. *J. Urol.* Jan 1997;157(1):199-202; discussion 202-193.

[22] Scattoni V, Raber M, Abdollah F, et al. Biopsy schemes with the fewest cores for detecting 95% of the prostate cancers detected by a 24-core biopsy. *Eur. Urol.* Jan 2010;57(1):1-8.

[23] Babaian RJ, Toi A, Kamoi K, et al. A comparative analysis of sextant and an extended 11-core multisite directed biopsy strategy. *J. Urol.* Jan 2000;163(1):152-157.

[24] Stewart CS, Leibovich BC, Weaver AL, Lieber MM. Prostate cancer diagnosis using a saturation needle biopsy technique after previous negative sextant biopsies. *J. Urol.* Jul 2001;166(1):86-91; discussion 91-82.

[25] Onik G, Miessau M, Bostwick DG. Three-dimensional prostate mapping biopsy has a potentially significant impact on prostate cancer management. *J. Clin. Oncol.* Sep 10 2009;27(26):4321-4326.

[26] Barzell WE, Melamed MR. Appropriate patient selection in the focal treatment of prostate cancer: the role of transperineal 3-dimensional pathologic mapping of the prostate--a 4-year experience. *Urology.* Dec 2007;70(6 Suppl):27-35.

[27] Ellis WJ, Brawer MK. The significance of isoechoic prostatic carcinoma. *J. Urol.* Dec 1994;152(6 Pt 2):2304-2307.

[28] Remzi M, Dobrovits M, Reissigl A, et al. Can Power Doppler enhanced transrectal ultrasound guided biopsy improve prostate cancer detection on first and repeat prostate biopsy? *Eur. Urol.* Oct 2004;46(4):451-456.

[29] Kimura G, Nishimura T, Kimata R, Saito Y, Yoshida K. Random systematic sextant biopsy versus power doppler ultrasound-guided target biopsy in the diagnosis of prostate cancer: positive rate and clinicopathological features. *J. Nihon Med. Sch.* Oct 2005;72(5):262-269.

[30] Yang JC, Tang J, Li J, Luo Y, Li Y, Shi H. Contrast-enhanced gray-scale transrectal ultrasound-guided prostate biopsy in men with elevated serum prostate-specific antigen levels. *Acad. Radiol.* Oct 2008;15(10):1291-1297.

[31] Taverna G, Morandi G, Seveso M, et al. Colour Doppler and microbubble contrast agent ultrasonography do not improve cancer detection rate in transrectal systematic prostate biopsy sampling. *BJU Int.* Dec 2011;108(11):1723-1727.

[32] Colleselli D, Bektic J, Schaefer G, et al. The influence of prostate volume on prostate cancer detection using a combined approach of contrast-enhanced ultrasonography-targeted and systematic grey-scale biopsy. *BJU Int.* Dec 2007;100(6):1264-1267.

[33] Kravchick S, Cytron S, Peled R, London D, Sibi Y, Ben-Dor D. Optimal combinations for detection of prostate cancer: systematic sextant and laterally directed biopsies versus systematic sextant and color Doppler-targeted biopsies. *Urology.* Feb 2004;63(2):301-305.

[34] Mitterberger M, Horninger W, Pelzer A, et al. A prospective randomized trial comparing contrast-enhanced targeted versus systematic ultrasound guided biopsies: impact on prostate cancer detection. *Prostate.* Oct 1 2007;67(14):1537-1542.

[35] Frauscher F, Klauser A, Volgger H, et al. Comparison of contrast enhanced color Doppler targeted biopsy with conventional systematic biopsy: impact on prostate cancer detection. *J. Urol.* Apr 2002;167(4):1648-1652.

[36] Pelzer A, Bektic J, Berger AP, et al. Prostate cancer detection in men with prostate specific antigen 4 to 10 ng/ml using a combined approach of contrast enhanced color Doppler targeted and systematic biopsy. *J. Urol.* Jun 2005;173(6):1926-1929.

[37] Mitterberger M, Pinggera G, Horninger W, et al. Dutasteride prior to contrast-enhanced colour Doppler ultrasound prostate biopsy increases prostate cancer detection. *Eur. Urol.* Jan 2008;53(1):112-117.

[38] Morelli G, Pagni R, Mariani C, et al. Results of vardenafil mediated power Doppler ultrasound, contrast enhanced ultrasound and systematic random biopsies to detect prostate cancer. *J. Urol.* Jun 2011;185(6):2126-2131.

[39] Ophir J, Cespedes I, Ponnekanti H, Yazdi Y, Li X. Elastography: a quantitative method for imaging the elasticity of biological tissues. *Ultrason Imaging.* Apr 1991;13(2):111-134.

[40] Krouskop TA, Wheeler TM, Kallel F, Garra BS, Hall T. Elastic moduli of breast and prostate tissues under compression. *Ultrason Imaging.* Oct 1998;20(4):260-274.

[41] Sumura M, Shigeno K, Hyuga T, Yoneda T, Shiina H, Igawa M. Initial evaluation of prostate cancer with real-time elastography based on step-section pathologic analysis after radical prostatectomy: a preliminary study. *Int. J. Urol.* Sep 2007;14(9):811-816.

[42] Cochlin DL, Ganatra RH, Griffiths DF. Elastography in the detection of prostatic cancer. *Clin. Radiol.* Nov 2002;57(11):1014-1020.

[43] Konig K, Scheipers U, Pesavento A, Lorenz A, Ermert H, Senge T. Initial experiences with real-time elastography guided biopsies of the prostate. *J. Urol.* Jul 2005;174(1):115-117.

[44] Pallwein L, Mitterberger M, Pinggera G, et al. Sonoelastography of the prostate: comparison with systematic biopsy findings in 492 patients. *Eur. J. Radiol.* Feb 2008;65(2):304-310.

[45] Brock M, von Bodman C, Palisaar RJ, et al. The impact of real-time elastography guiding a systematic prostate biopsy to improve cancer detection rate: a prospective study of 353 patients. *J. Urol.* Jun 2012;187(6):2039-2043.

[46] Pallwein L, Aigner F, Faschingbauer R, et al. Prostate cancer diagnosis: value of real-time elastography. *Abdom Imaging.* Nov-Dec 2008;33(6):729-735.

[47] Heijmink SW, Barentsz JO. Contrast-enhanced versus systematic transrectal ultrasound-guided prostate cancer detection: an overview of techniques and a systematic review. *Eur. J. Radiol.* Sep 2007;63(3):310-316.

[48] Bigler SA, Deering RE, Brawer MK. Comparison of microscopic vascularity in benign and malignant prostate tissue. *Hum. Pathol.* Feb 1993;24(2):220-226.

[49] Tang J, Yang JC, Luo Y, Li J, Li Y, Shi H. Enhancement characteristics of benign and malignant focal peripheral nodules in the peripheral zone of the prostate gland studied using contrast-enhanced transrectal ultrasound. *Clin. Radiol.* Oct 2008;63(10):1086-1091.

[50] Halpern EJ, Frauscher F, Rosenberg M, Gomella LG. Directed biopsy during contrast-enhanced sonography of the prostate. *AJR Am. J. Roentgenol.* Apr 2002;178(4):915-919.

[51] Linden RA, Trabulsi EJ, Forsberg F, Gittens PR, Gomella LG, Halpern EJ. Contrast enhanced ultrasound flash replenishment method for directed prostate biopsies. *J. Urol.* Dec 2007;178(6):2354-2358.

[52] Aigner F, Pallwein L, Mitterberger M, et al. Contrast-enhanced ultrasonography using cadence-contrast pulse sequencing technology for targeted biopsy of the prostate. *BJU Int.* Feb 2009;103(4):458-463.

[53] Simmons LA, Autier P, Zat'ura F, et al. Detection, localisation and characterisation of prostate cancer by prostate HistoScanning(). *BJU Int.* Jul 2012;110(1):28-35.

[54] Huynen AL, Giesen RJ, de la Rosette JJ, Aarnink RG, Debruyne FM, Wijkstra H. Analysis of ultrasonographic prostate images for the detection of prostatic carcinoma: the automated urologic diagnostic expert system. *Ultrasound Med. Biol.* 1994;20(1):1-10.

[55] Giesen RJ, Huynen AL, Aarnink RG, et al. Computer analysis of transrectal ultrasound images of the prostate for the detection of carcinoma: a prospective study in radical prostatectomy specimens. *J. Urol.* Oct 1995;154(4):1397-1400.

[56] Beerlage HP, Aarnink RG, Ruijter ET, et al. Correlation of transrectal ultrasound, computer analysis of transrectal ultrasound and histopathology of radical prostatectomy specimen. *Prostate Cancer Prostatic Dis.* 2001;4(1):56-62.

[57] Grabski B, Baeurle L, Loch A, Wefer B, Paul U, Loch T. Computerized transrectal ultrasound of the prostate in a multicenter setup (C-TRUS-MS): detection of cancer after multiple negative systematic random and in primary biopsies. *World J. Urol.* Oct 2011;29(5):573-579.

[58] Prager RW, Ijaz UZ, Gee AH, Treece GM. Three-dimensional ultrasound imaging. *Proc. Inst. Mech. Eng. H.* 2010;224(2):193-223.

[59] Elliot TL, Downey DB, Tong S, McLean CA, Fenster A. Accuracy of prostate volume measurements in vitro using three-dimensional ultrasound. *Acad. Radiol.* May 1996;3(5):401-406.

[60] Tong S, Downey DB, Cardinal HN, Fenster A. A three-dimensional ultrasound prostate imaging system. *Ultrasound Med. Biol.* 1996;22(6):735-746.

[61] Shen F, Shinohara K, Kumar D, et al. Three-dimensional sonography with needle tracking: role in diagnosis and treatment of prostate cancer. *J. Ultrasound Med.* Jun 2008;27(6):895-905.

[62] Long JA, Daanen V, Moreau-Gaudry A, Troccaz J, Rambeaud JJ, Descotes JL. Prostate biopsies guided by three-dimensional real-time (4-D) transrectal ultrasonography on a phantom: comparative study versus two-dimensional transrectal ultrasound-guided biopsies. *Eur. Urol.* Oct 2007;52(4):1097-1104.

[63] Abul FT, Arun N, Abu-Assi MA, Asbeutah AM. Transrectal ultrasound guided biopsy for detecting prostate cancer: can random biopsies be reduced using the 4-dimensional technique? *Int. Urol. Nephrol.* 2007;39(2):517-524.

[64] Cool DW, Connolly MJ, Sherebrin S, et al. Repeat prostate biopsy accuracy: simulator-based comparison of two- and three-dimensional transrectal US modalities. *Radiology.* Feb 2010;254(2):587-594.

[65] Bax J, Cool D, Gardi L, et al. Mechanically assisted 3D ultrasound guided prostate biopsy system. *Med Phys.* Dec 2008;35(12):5397-5410.

[66] Narayanan R, Werahera PN, Barqawi A, et al. Adaptation of a 3D prostate cancer atlas for transrectal ultrasound guided target-specific biopsy. *Phys. Med. Biol.* Oct 21 2008;53(20):N397-406.

[67] Schuster D, Fei B, Fox T, Osunkoya AO. Histopathologic Correlation of Prostatic Adenocarcinoma on Radical Prostatectomy with Pre-Operative Anti-18F Fluorocyclobutyl-Carboxylic Acid Positron Emission Tomography/Computed Tomography. *Lab. Invest.* Feb 2011;91:222a-223a.

[68] Fei B, Schuster DM, Master V, Akbari H, Fenster A, Nieh P. A Molecular Image-directed, 3D Ultrasound-guided Biopsy System for the Prostate. *Proc SPIE.* 2012;2012.

[69] Esenaliev RO, Karabutov AA, Oraevsky AA. Sensitivity of laser opto-acoustic imaging in detection of small deeply embedded tumors. *Ieee J. Sel Top Quant.* Jul-Aug 1999;5(4):981-988.

[70] Yaseen MA, Ermilov SA, Brecht HP, et al. Optoacoustic imaging of the prostate: development toward image-guided biopsy. *J. Biomed. Opt.* Mar-Apr 2010;15(2):021310.

[71] Wang K, Ermilov SA, Su R, Brecht HP, Oraevsky AA, Anastasio MA. An imaging model incorporating ultrasonic transducer properties for three-dimensional optoacoustic tomography. *IEEE Trans. Med. Imaging.* Feb 2011;30(2):203-214.

[72] Kirkham AP, Emberton M, Allen C. How good is MRI at detecting and characterising cancer within the prostate? *Eur. Urol.* Dec 2006;50(6):1163-1174; discussion 1175.

[73] Engelbrecht MR, Jager GJ, Laheij RJ, Verbeek AL, van Lier HJ, Barentsz JO. Local staging of prostate cancer using magnetic resonance imaging: a meta-analysis. *Eur. Radiol.* Sep 2002;12(9):2294-2302.

[74] Wefer AE, Hricak H, Vigneron DB, et al. Sextant localization of prostate cancer: comparison of sextant biopsy, magnetic resonance imaging and magnetic resonance spectroscopic imaging with step section histology. *J. Urol.* Aug 2000;164(2):400-404.

[75] Qayyum A. Diffusion-weighted imaging in the abdomen and pelvis: concepts and applications. *Radiographics.* Oct 2009;29(6):1797-1810.

[76] Chen YJ, Chu WC, Pu YS, Chueh SC, Shun CT, Tseng WY. Washout gradient in dynamic contrast-enhanced MRI is associated with tumor aggressiveness of prostate cancer. *J. Magn Reson Imaging.* Jun 18 2012.

[77] Kobus T, Vos PC, Hambrock T, et al. Prostate Cancer Aggressiveness: In Vivo Assessment of MR Spectroscopy and Diffusion-weighted Imaging at 3 T. *Radiology.* Jul 27 2012.

[78] D'Amico AV, Tempany CM, Cormack R, et al. Transperineal magnetic resonance image guided prostate biopsy. *J. Urol.* Aug 2000;164(2):385-387.

[79] Ewertsen C, Grossjohann HS, Nielsen KR, Torp-Pedersen S, Nielsen MB. Biopsy guided by real-time sonography fused with MRI: a phantom study. *AJR Am. J. Roentgenol.* Jun 2008;190(6):1671-1674.

[80] Xu S, Kruecker J, Turkbey B, et al. Real-time MRI-TRUS fusion for guidance of targeted prostate biopsies. *Comput Aided Surg.* Sep 2008;13(5):255-264.

[81] Kuru TH, Roethke M, Popeneciu V, et al. Phantom study of a novel stereotactic prostate biopsy system integrating preinterventional magnetic resonance imaging and live ultrasonography fusion. *J. Endourol.* Jul 2012;26(7):807-813.

[82] Singh AK, Kruecker J, Xu S, et al. Initial clinical experience with real-time transrectal ultrasonography-magnetic resonance imaging fusion-guided prostate biopsy. *BJU Int.* Apr 2008;101(7):841-845.

[83] Kaplan I, Oldenburg NE, Meskell P, Blake M, Church P, Holupka EJ. Real time MRI-ultrasound image guided stereotactic prostate biopsy. *Magn Reson Imaging.* Apr 2002;20(3):295-299.

[84] Ukimura O, Hirahara N, Fujihara A, et al. Technique for a hybrid system of real-time transrectal ultrasound with preoperative magnetic resonance imaging in the guidance of targeted prostate biopsy. *Int. J. Urol.* Oct 2010;17(10):890-893.

[85] Hadaschik BA, Kuru TH, Tulea C, et al. A novel stereotactic prostate biopsy system integrating pre-interventional magnetic resonance imaging and live ultrasound fusion. *J. Urol.* Dec 2011;186(6):2214-2220.

[86] Pinto PA, Chung PH, Rastinehad AR, et al. Magnetic resonance imaging/ultrasound fusion guided prostate biopsy improves cancer detection following transrectal ultrasound biopsy and correlates with multiparametric magnetic resonance imaging. *J. Urol.* Oct 2011;186(4):1281-1285.
[87] Rud E, Baco E, Eggesbo HB. MRI and Ultrasound-guided Prostate Biopsy Using Soft Image Fusion. *Anticancer Res.* Aug 2012;32(8):3383-3389.
[88] Natarajan S, Marks LS, Margolis DJ, et al. Clinical application of a 3D ultrasound-guided prostate biopsy system. *Urol Oncol.* May-Jun 2011;29(3):334-342.
[89] Turkbey B, Xu S, Kruecker J, et al. Documenting the location of prostate biopsies with image fusion. *BJU Int.* Jan 2011;107(1):53-57.
[90] Smeenge M, de la Rosette JJ, Wijkstra H. Current status of transrectal ultrasound techniques in prostate cancer. *Curr. Opin Urol.* May 16 2012.

In: Ultrasonography
Editor: Alexander Burdenko
ISBN: 978-1-62417-536-7

Chapter 2

ULTRASOUND IN THE MANAGEMENT OF ISCHEMIC STROKE

***Narayanaswamy Venketasubramanian*[1,2]* *and Yohanna Kusuma*[2]**
[1]Neuroscience Clinic, Raffles Hospital, Singapore
[2]Division of Neurology, University Medical Cluster, National University Health System, National University of Singapore, Singapore

ABSTRACT

Ultrasound is increasingly being used in the diagnosis, management and treatment of ischemic stroke. Linear array probes with frequencies of 5 to 10 MHz are used for extracranial studies, and sector or small hand-held 2 MHz probes for transcranial studies. Ultrasound is non-invasive, safe, painless, portable, quick, inexpensive, repeatable and accurate, but it requires adequate training; accreditation for physicians and laboratories are available. The extracranial arteries imaged include the common, internal and external carotid, vertebral, subclavian arteries bilaterally, and the brachiocephalic trunk. For transcranial studies, the ophthalmic artery, carotid siphon, terminal internal carotid, and proximal anterior, middle, posterior, basilar and intracranial vertebral arteries are evaluated. Brightness-(B)mode, intensity/power mode, colour-coded and Doppler

* Corresponding author: Narayanaswamy Venketasubramanian. E-mail: drnvramani@gmail.com.

mode are used; more recently, motion-(M)mode is also employed. Velocity parameters measured include the Peak Systolic, End-Diastolic and Mean velocities; resistivity parameters include the Gosling's Pulsatility Index and Pourcelot's Resistance Index. For safety, the Thermal Index cannot exceed 6, Mechanical Index 1.9, Spatial Peak Temporal Average Intensity 720 mW/cm2. Ultrasound is used to evaluate extracranial and intracranial arterial stenosis, collateral flow as a result of severe stenosis/ occlusion, pulsatile neck masses, subclavian steal, carotid dissection, Takayasu disease, right-to-left shunting, embolisation, intra-operative monitoring during carotid endarterectomy or CABG and post-surgery/stent assessment of stenosis, as well as cerebrovascular reserve to predict stroke risk. It is also used therapeutically to enhance thrombolysis (sonothrombolysis). Neurosonology provides an invaluable tool for the clinician in the management and prevention of ischemic stroke.

Introduction

Stroke is a leading cause of death and disability in many parts of the world. With the increasing numbers and aging of the population, as well as economic transition of less developed to more developed status, the number of people suffering a stroke will likely rise [1]. Approximately 54 – 85% of strokes are ischaemic in nature, with varying mechanisms including large artery atherosclerosis (includes atherothromobosis, artery-to-artery embolism and hypoperfusion), small artery occlusion, cardioembolism, other known causes, or unknown [2]. Large artery disease is largely amenable to detection by ultrasonography.

Why Ultrasound

Ultrasonography is widely used in many fields of Medicine. It is increasingly being used in the diagnosis and management of ischaemic stroke. It is non-invasive, safe, painless, portable, quick, inexpensive, repeatable and reasonably accurate compared to other imaging modalities [3]. Computed tomographic angiography(CTA) and magnetic resonance angiography (MRA) should be considered as complementary rather than competing techniques. CTA requires the availability of a higher-end CT scanner, exposure to radiation and the injection of a contrast medium. MRA requires the availability of an MR scanner, is contraindicated in those with metallic implants or pacemakers, is a challenge in claustrophic or very obese individuals or those who are haemodynamically unstable or on ventilatory support or restless, and has a long imaging time. The 'gold standard' remains catheter angiography with its antecedent risks of radiation, peri-procedural stroke, need for contrast injection, invasiveness and the need for expertise to perform it. None of these techniques is portable; they do not give real-time information; doing frequent, repeated studies and studies at the bedside is thus not possible or feasible. Cost-wise they are far more expensive than ultrasound, in terms of equipment purchase and maintainence and operating costs (Table 1).

Table 1. Comparison of neurovascular imaging methods

Features	Ultrasound	Computed Tomography Angiogram	Magnetic Resonance Angiogram	Catheter Angiography
Accuracy	++	+++	++	Gold standard
Low cost	+++	++	++	+
Speed	++	+++	++	+
Portability	++	0	0	0
Non-invasiveness	+++	+++	+++	0
Lack of radiation	+++	0	+++	0
Lack of contra-indications	+++	++	+	+
Low risk	+++	++	++	+

Clinical Cerebrovascular Anatomy and Physiology

The brain receives blood supply containing oxygen and the only substrate it can use glucose via the carotid and vertebral arterial systems. The right common carotid artery (CCA) is a terminal branch of the brachiocephalic trunk (BCT, also called the innominate artery) which rises from the aortic arch in the thorax; the other terminal branch is the right subclavian artery (ScA). The left CCA and ScA arise directly and independently from the aortic arch. The diameter of the CCA is approximately 6mm.

The jugular vein on each side often lies superficial to its respective CCA, and can be identified by its easy compressibility, presence of valves, and flow in the opposite direction to the CCA. Each CCA ascends cephalically and bifurcates in the mid-neck usually at the level of the thyroid cartilage into the internal (ICA) and external (ECA) carotid arteries. The ICA then enters the petrous temporal bone.

The ICA can be differentiated from the ECA by a number of features – posterior-lateral location, dilatation at origin, larger diameter, low resistance flow pattern; the lack of branches is the key differentiating feature as is the Doppler waveform response to superficial temporal artery tapping (Table 2).

Table 2. Differentiating internal and external carotid arteries by ultrasonography

Feature	Internal Carotid Artery	External Carotid Artery
Location	Posterolateral	Anteromedial
Dilation at Origin	Yes	No
Diameter	Larger	Smaller
Branches in the neck	No	Yes
Flow pattern	Low resistance	High resistance
Response to superficial temporal artery tapping	No	Yes

The first branch of the ECA is usually the superior thyroid artery. The ECA's branches form many anastomoses with the ICA – superficial temporal artery with the supraorbital artery which is a branch of the ophthalmic artery (OA); facial artery with the dorsal nasal artery which is another branch of the OA.

The vertebral artery (VA) arises from the second part of the ScA and runs medially to enter the transverse process of the sixth vertebra. Each VA then ascends cephalically in the transverse foramen where it is easily and reliably identified, before it exits after traversing the second vertebra, winds round the transverse process of the atlas and enters the foramen magnum. The diameter of each VA is 2-3 mm, and may be dominant in one side, usually the right.

In its course through the petrous temporal bone, the ICA enters the cavernous sinus, runs anteriorly and then loops back, forming the carotid siphon (CS) where the OA is given off – OA flow is towards the eye. The ICA then again ascends cephalically, gives off the posterior communicating artery (PCommA) which joins the posterior cerebral artery (PCA), and terminates as a bifurcation into the middle cerebral artery (MCA) and anterior cerebral artery (ACA).

The MCA runs laterally (M1 segment) before bifurcating or even trifurcating in the Sylvian fissure (M2 segments). The ACA runs medially (A1) and has a branch to the other ACA called the anterior communicating artery (ACommA), after which it becomes the A2 segment.

The VA in its intracranial portion gives off the posterior inferior cerebellar artery (PICA) and then joins the other VA at the ponto-medullary junction to form the basilar artery (BA). The BA runs cephalically over the pons and terminates at the midbrain-pons junction as the two PCAs. Each PCA runs laterally (P1 segment), is joined by the PCommA after which it becomes the P2 segment and then curves and runs posteriorly.

The Circle of Willis lies at the base of the brain and runs around the pituitary gland. It comprises the distal intracranial ICAs, M1s, A1s, ACommA, PCommAs and P1s, an important intracranial collateral circulation that is complete in only 50% of people.

Approximately 15% of the cardiac output, or about 750 ml/min of blood, is supplied to the adult brain, or 50-55 ml/100g brain/min – it is 75 ml/100g/min in grey matter and 45 ml/100g/min in white matter. Each CCA carries approximately 300-350 ml/min, each VA 50-100 ml/min. Cerebral autoregulation maintains blood flow within a constant range at systemic blood pressures between 50 and 150 mmHg. This is largely achieved by vasodilation and vasoconstriction of cerebral arterioles. The first response to reduced cerebral perfusion is cerebral vasodilation. In severe ischaemia, vasodilation is maximal; the inability for further dilation is referred to as a loss of cerebrovascular reserve, a portend of further ischaemic cell injury and stroke due to hypoperfusion.

Imaging Modalities, Instrumentation and Indices

Extracranial Imaging usually studies the CCAs including the RCCA origin off the BCT, carotid bifurcations, proximal and mid-ICAs and ECAs, and proximal and vertebral portions of the VAs including their origins off the ScAs. The structural integrity and anatomy of the vessels, intimamedial thickness (IMT), plaques, intimal flaps, and luminal thrombi are imaged using B- (= Brightness) mode. Intensity-, also called power-, mode imaging detects the presence of flow. Colour-coded Doppler imaging gives information of the direction of flow and a qualitative estimate of velocities of moving red blood cells. Doppler imaging transforms frequency shifts to velocities; pulse-wave Doppler is preferred over continuous-wave Doppler as specific depths can be targeted and interrogated. The combination of B-mode and Doppler is called Duplex Imaging; the addition of colour-coding makes it Triplex Imaging.

Due to poor resolution, B-mode is used in Transcranial Imaging (TCI) only to locate the midbrain after which Triplex Imaging is performed to locate arteries and determine flow direction and velocities. Intensity-, also called power-, mode imaging is also used. Transcranial Doppler (TCD) is a blind technique using only Doppler to detect arteries and determine flow direction and velocity at various depths. B-mode has been used to detect large intracranial hemorrhages and hydrocephalus, but CT and MRI are far superior techniques. TCD has also been used to detect high intensity transient signal (HITs) presumed to be due to microemboli and to assess intracranial flow changes after manoeuvres. M- (=Motion) mode is a new TCD technique that allows multiple depths to be studied simultaneously. In addition to detecting areas of stenosis, TCI and TCD also detect the intracranial effects of extracranial disease including collateral flow, and TCD is used to detect embolisation from a more proximal source, and assessment of cerebrovascular reserve. They are also used therapeutically to enhance thrombolysis (sonothrombolysis).

Doppler frequencies of 5 to 10 MHz are used for extracranial imaging as the arteries are superficial; linear array probes are used, with higher frequencies allowing better spatial resolution. Transcranial imaging requires 2 to 2.5 MHz frequencies due to the attenuation by the bones of the skull and deeper location of structures to be imaged; sector or small hand-held probes are used. Duplex and Triplex machines are large and may be portable if wheel-mounted; laptop and even hand-held versions are now available. Pure TCD machines are smaller, and may even be small enough to be carried in a back-pack.

Velocity parameters measured include the peak systolic (PSV), end-diastolic (EDV) and mean (MV) [= EDV + (PSV-EDV)/3] velocities. Resistivity parameters include the Gosling's Pulsatility Index [= (PSV-EDV)/MV] and Pourcelot's Resistance Index (RI) [= (PSV-EDV)/PSV]. A number of factors affect velocities, including age, heart rate, blood pressure, haematocrit, fever, hypoglycemia, and hypercarbia.

SAFETY

Ultrasound may potentially harm body tissues due to the high intensity of ultrasound energy used and prolonged duration of insonation, causing thermal (heating) and non-thermal (cavitation, streaming) effects. Heating affects the integrity of tissues and proteins, and the rate and equilibrium of chemical reactions. The Thermal Index (TI) is the regulated matrix; the TI in Soft tissue (TIS) and TI in the Cranium (TIC) are the sub-indices of relevance to neurosonology. Cavitation is due to the expansion and contraction (stable cavitation), and the more dangerous collapse (transient cavitation) of gas bubbles during the oscillatory cycle; pressure, force, torque and streaming are due to the force of pressures. The primary regulated metric is the Mechanical Index (MI). Currently the US Food and Drug Administration stipulates that diagnostic ultrasound scanners cannot exceed a TI of 6 and an MI of 1.9; the Spatial Peak Temporal Average Intensity I(SPTA) cannot exceed 720 mW/cm2. For ophthalmic imaging, I(SPTA) cannot exceed 17 mW/cm2 while MI should be less that 0.23. At all times, the As Low As Reasonably Achievable (ALARA) principle should be used – the lowest energy intensity and shortest study duration that allows a reasonable study [4].

PERFORMING AN EXTRACRANIAL TRIPLEX EXAMINATION

The patient is placed supine with the head partially laterally rotated away from the side to be examined. The gel-covered probe in B-mode is gently placed transversely in the mid-neck over the sternomastoid muscle and the CCA imaged as a 1cm diameter pulsatile structure with a dark (=anechoic) lumen and bright (=echoic) wall (Figure 1). It is then traced proximally to the base of the neck; on the right ride, the BCT is also imaged. The probe is then moved cephalically and the carotid bifurcation identified as two circular pulsatile structures (Figure 2). The ICA and ECA are differentiated and the ICA traced as distally as possible. The probe is then placed at the bifurcation and rotated so as to image the bifurcation in the longitudinal plane (Figure 3). The distal CCA, proximal ICA and ECA are identified. The vessels are traced proximally and distally. The intimal-media thickness (IMT) is measured, usually in the far wall of the distal CCA (Figure 4). Then the intensity/power mode is turned on to confirm flow in all segments of the arteries (Figure 5). The colour-coded Doppler is then used to confirm flow direction and identify areas of turbulence which may represent sites of stenosis (Figure 6). Finally, in the Triplex mode, the Doppler is used with a narrow sample volume placed in the centre of the artery at specific points as well as areas of turbulence, and an insonation angle steered along the direction of flow, not to be more than 60 degrees (Figure 7). The highest velocities are recorded in the proximal and distal CCA; origin, bulb and as distally in the ICA as possible; proximal ECA.

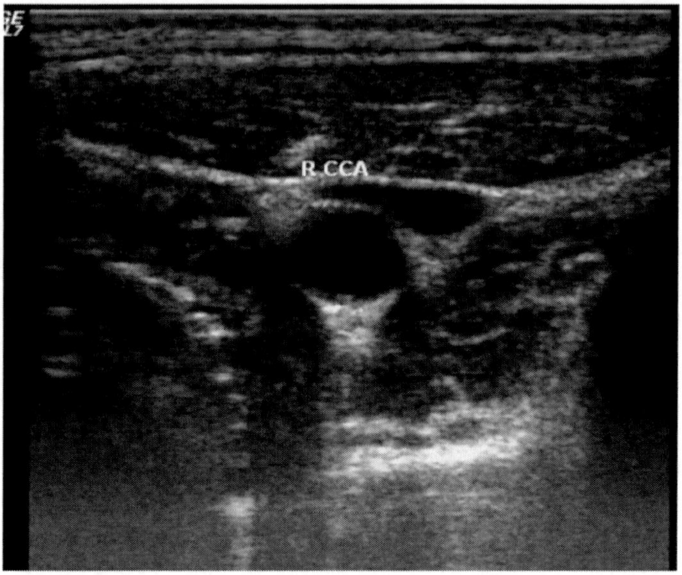

Figure 1. Right common carotid artery, cross-section, B-mode.

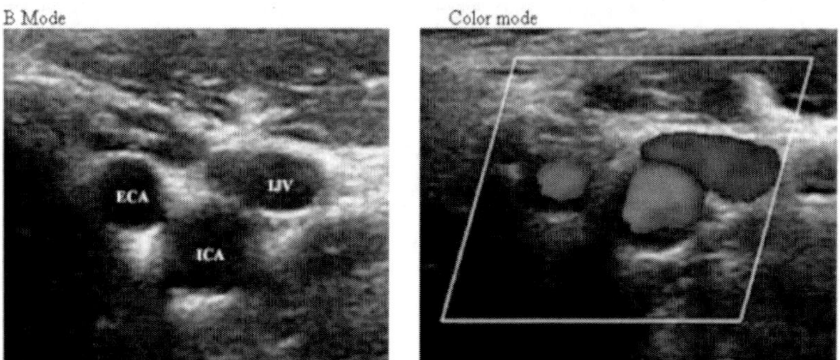

Figure 2. Internal carotid artery, external carotid artery, internal jugular vein, cross-section B-mode (left), colour-coded Doppler mode (right).

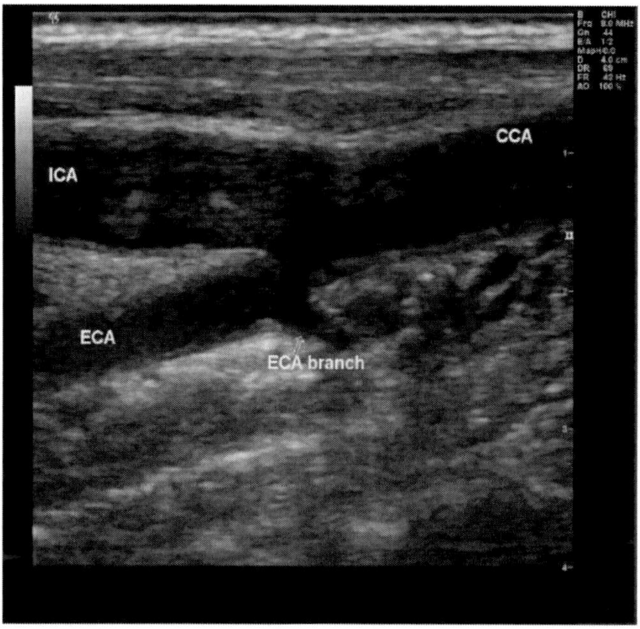

Figure 3. Carotid bifurcation, longitudinal section, B-mode.

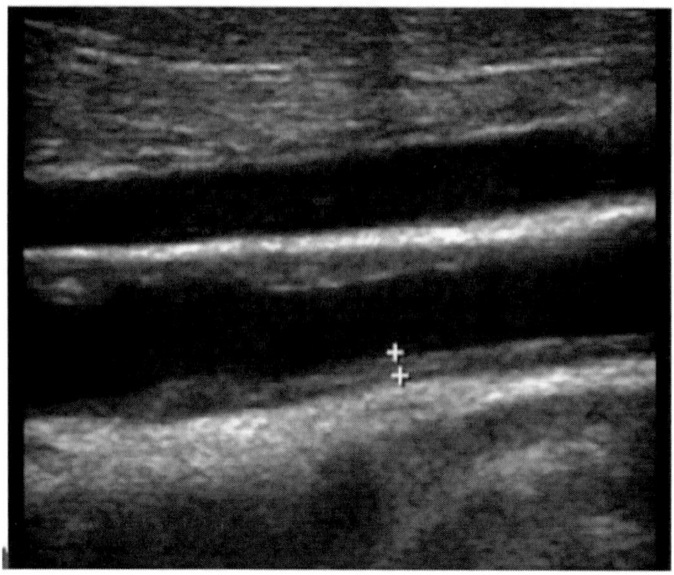

Figure 4. Measurement of intimamedial thickness in far wall of mid-common carotid artery.

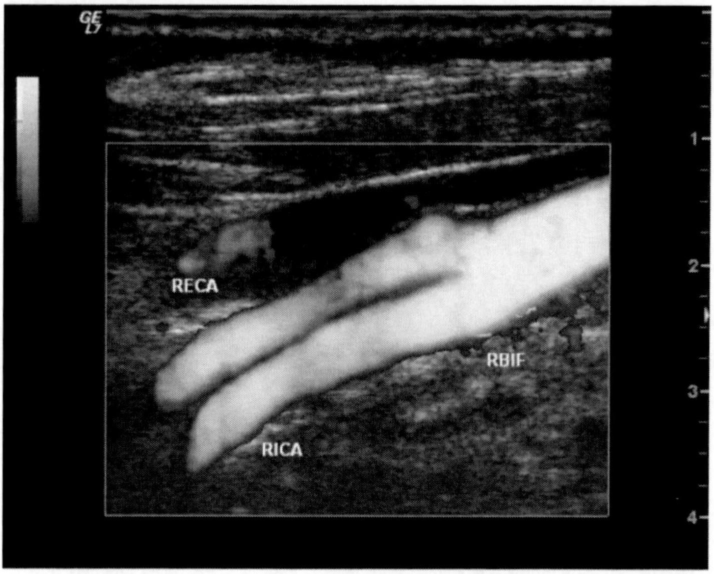

Figure 5. Right carotid bifurcation, longitudinal section, power/intensity mode.

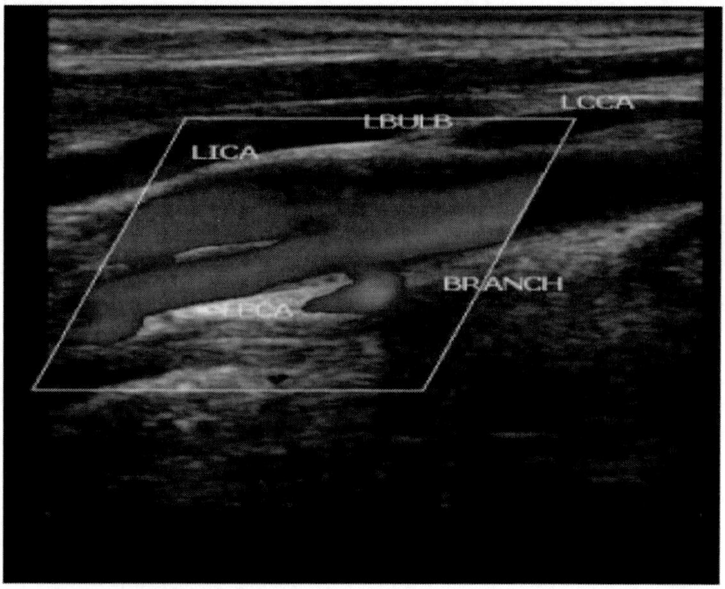

Figure 6. Left carotid bifurcation, longitudinal section, colour-coded Doppler mode.

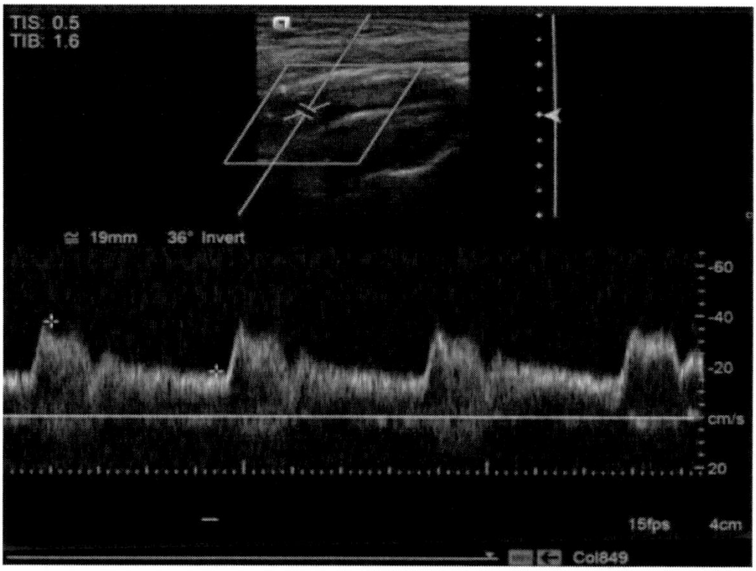

Figure 7. Mid-internal carotid artery, longitudinal section, triplex mode.

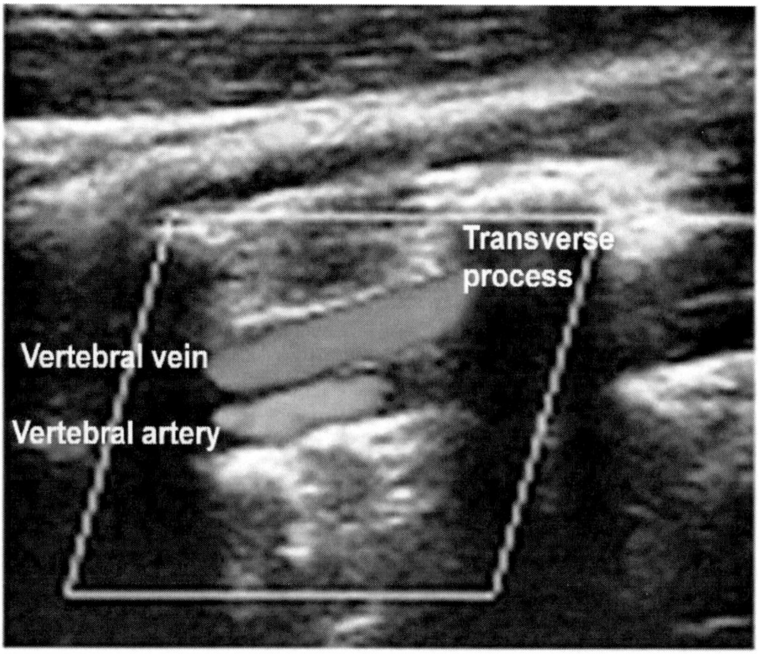

Figure 8. Vertebral artery, longitudinal section, colour-coded Doppler mode.

The B-mode only is then turned on and the probe held vertically over the mid-sternomastoid to identify the VA in its intervertebral segment as a linear structure between the transverse processes of the cervical spines (Figure 8).

The intensity/power, then color-coded Doppler modes are used to study the entire VA including its proximal portions and its origin from the ScA. The Doppler mode is used as for the carotid study, evaluating the mid- and proximal VA and VA origin, as well as areas of turbulence.

The head is then turned the other way and the above processes repeated to study the contralateral carotid and vertebral arterial systems.

Challenges to a successful study include a patient's short and thick neck, high carotid bifurcation, breathlessness, and repeated swallowing.

PERFORMING A TRANSCRANIAL TRIPLEX EXAMINATION

The patient is placed supine, and the gel-covered probe in B-mode is placed gently over the temporal region above the ear in the transverse plane. The probe may be moved back and forth in the temporal window looking for the butterfly-shaped hypoechoic midbrain. Once found, the colour-coded Doppler is turned on and the vessels of the circle of Willis identified (Figure 9). The PCA wraps around the cerebral peduncle. More anteriorly, the MCA travels towards the probe while the ACA travels away.

Often, the contralateral ACA and PCA, and even MCA, may be seen. Areas of turbulence are interrogated by Doppler with the sample volume covering the whole artery to detect areas of significant stenosis, recording maximum velocities in the MCA, ACA and PCA; angle correction is not usually performed. The contralateral side is then examined in a similar manner.

The patient's head is then turned laterally, or the patient is sat up, with the head bent forward. The probe in B-mode in transverse plane is then placed over midline of the nape of the neck posteriorly, just below the occipital nuchae. The foramen magnum is identified as a hypoechoic round structure. The colour-coded Doppler is turned on and the VAs are identified entering the foramen magnum (Figure 10).

At times, the probe has to be placed a little to the side to obtain good images. The VAs are tracked further intracranially where they merge to form the BA which travels further cephalically. The posterior inferior cerebellar artery (PICA) is often identified as a large branch of the intracranial VA. Doppler is used to detect areas of significant stenosis in the VAs and BA.

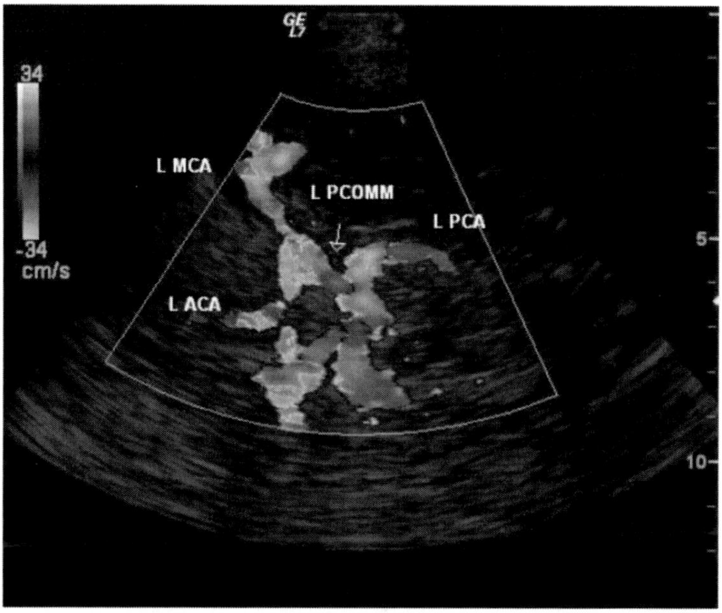

Figure 9. Circle of Willis, left temporal window, colour-coded Doppler mode.

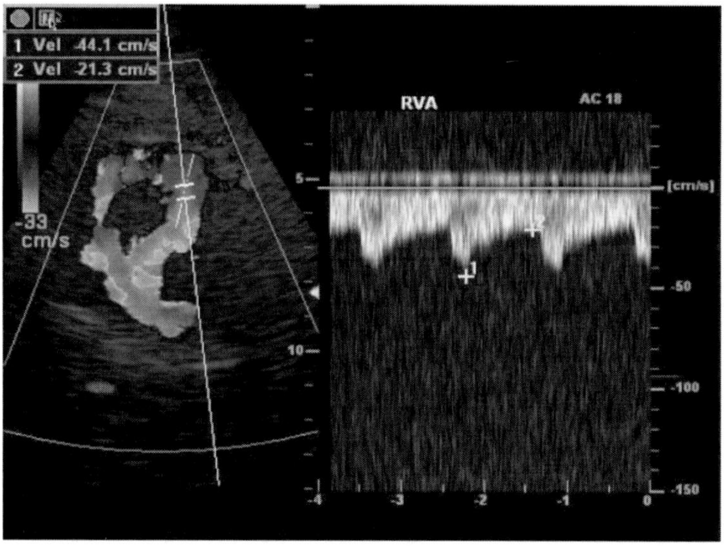

Figure 10. Vertebrobasilar junction, foraminal window, colour-coded Doppler mode.

With the power reduced to 10%, the lightly-gelled probe in transverse plane is gently placed over the eye ball and pointed slightly medially (= orbital window). The colour-coded Doppler is used detect the OA which flows towards the probe, and Doppler used to determine velocities.

At greater depths, the CS is detected with flow towards then away from the probe. OA and CS velocities are recorded. The contralateral side is then examined in a similar manner.

The challenges include the absence of a temporal window on one or both sides in up to 30% of patients, especially those who are elderly, female or of Asian ethnicity. A restless patient may not allow a detailed examination to be performed.

Performing a Transcranial Doppler (TCD) Examination

The patient is positioned as for Trancranial Triplex examination. The gel-tipped small hand-held Doppler probe is placed over the temporal window and pointed inwards and a little anteriorly. The probe is manipulated at various angles to pick up the arterial signal of the MCA. The signal is traced laterally by varying the depth of the sample volume, then again medially till the ICA bifurcation is reached, recognised by MCA flows towards and ACA flows away from the probe. Tilting the probe inferiorly insonates the terminal ICA. The probe is then turned posteriorly and PCA examined. The contralateral arteries are examined in a similar manner.

The probe is then placed over the nape of the neck and the VAs and BA studied by tracking the VA signal through the foraminal window (Figure 11). Finally, with the power reduced to 10%, the probe is placed over the orbital window to investigate OA and CS. Both sides are examined.

The probe may also be placed below the mandible and the distal extracranial ICA insonated on each side in turn.

As TCD is a blind technique, vessels are identified by the window through which they are detected, orientation of probe, depth of insonation, and direction of flow (Table 3). Other challenges are as for transcranial triplex examination, including the absence of a temporal window on one or both sides, and the restless patient.

Table 3. Identifying intracranial arteries by transcranial Doppler

Window	Artery	Orientation of probe	Depth of insonation (mm)	Direction of flow	Mean velocities (cm/s)
Temporal	MCA	Slightly anteriorly	40 – 65	Towards probe	60
	ACA	Even more anteriorly	65-80	Away	50
	tICA	Inferiorly	60-70	Towards	40
	PCA	Posteriorly	55-80	Towards (P1) Away (P2)	40
Orbital	OA	Slightly medially	40-55	Towards	20 (high resistance)
	CS	Slightly medially	60-80	Towards or away	45
Foraminal	VA	If in midline, laterally If lateral, medially	40-90	Away	40
	BA	Midline towards nasion	80-100	Away	45
Submandibular	ICA	Towards ipsilateral eye	40-70	Away	40

Clinical Applications

Ultrasonography has a wide range of clinical applications in the management of ischaemic stroke

Intimamedial Thickness

Increasing intimamedial thickness (IMT) is associated with a moderately increased risk for cerebro- and cardiovascular events [5-10]. Statins statistically reduce the rate of intima-media thickness progression among middle-aged adults with raised LDL-cholesterol and moderate carotid IMT thickening; however, the clinical significance of this reduction is uncertain [11].

In general, IMT exceeding 1mm is considered thickened. IMT is usually measured along the far wall of the distal CCA just before the bifurcation using high-resolution B-mode imaging in the longitudinal section. The first marker is more superficial and is placed at the junction of the (dark) lumen and the (bright) intima. The second marker is placed more deep past the (bright) intima and (dark, presumed) media at the junction of the (dark, presumed) media and (bright) adventitia. Automated systems are available that will provide automatic estimations of IMT.

Extracranial Arterial Stenosis/Occlusion

Extracranial atherosclerosis typically occurs at the carotid bifurcation, possibly due to shear forces due to splitting of blood flows. Internal carotid plaque causes 10-20% of ischaemic strokes [12]. Plaques lead to cerebral ischemia by distal embolisation (of overlying thrombus, platelet clumps, lipid material), or haemodynamic effect of severe stenosis/occlusion leading to distal field ischaemia. Increasing carotid stenosis is associated with increasing risk of ipsilateral stroke and recurrent ichaemia (up to 26% at 2 years), which is reduced by early carotid endarterectomy [13].

Ultrasound is widely used as a screening tool for significant stenosis, and has even been employed as the primary screening tool for clinical trials [14]. Description of the plaque includes mention of the number of plaques, location, size, surface (smooth/ulcerated, the latter carrying higher risk of atheroembolism)), texture (homogenous/heterogenous), echogenicity (hypo-, iso-, hyper-echoic, with respect to the sternomastoid muscle, hypoechogenicity being of higher risk of plaque rupture), calcification (usually casting an acoustic shadow), and intraplaque haemorrhage (seen as a hypoechoic area within a plaque) (Figure 11).

There are a number of criteria that have been used to determine the degree of stenosis [15-17] (Figure 12).

Each neurovascular laboratory is encouraged to develop its own validated criteria for grading stenosis [18].

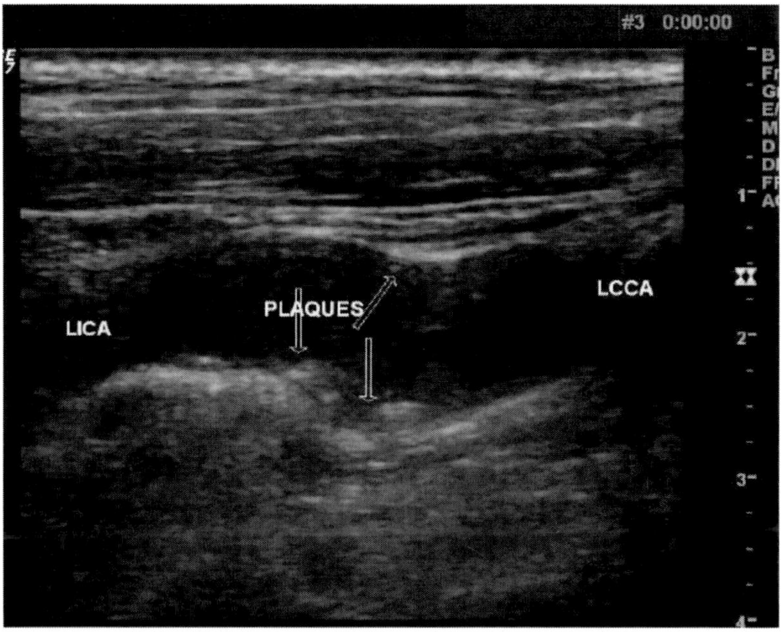

Figure 11. Left carotid bifurcation, longitudinal section, B-mode, showing plaques.

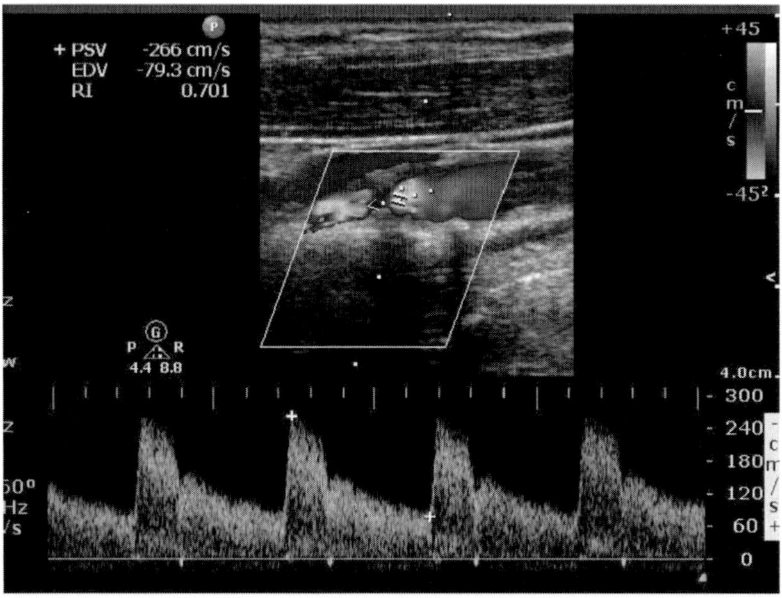

Figure 12. Proximal internal carotid artery, triplex mode, showing severe stenosis.

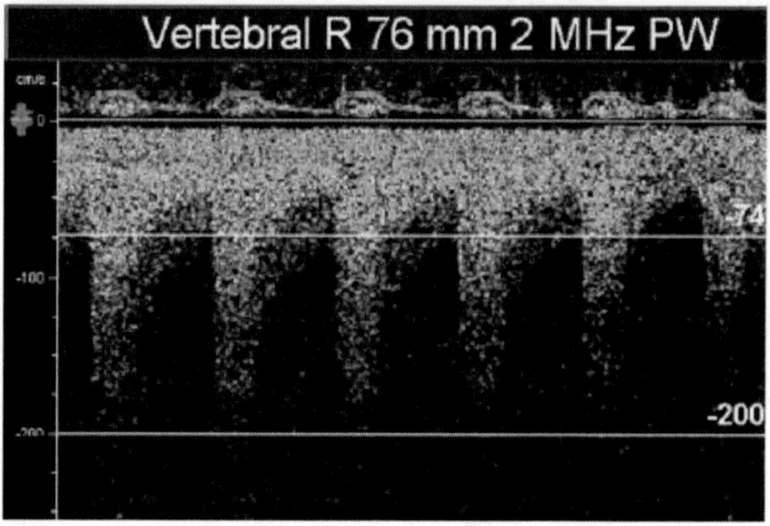

Figure 13. Right vertebral artery, foraminal window, transcranial Doppler, with high velocity showing stenosis.

Atherosclerotic stenosis may also occur at the origin of the VA, and less commonly along the mid-VA (Figure 13). The optimal management of VA stenosis (eg by stenting) is uncertain [19]. Diagnostic criteria for VA stenosis have been proposed [20].

Non-Atherosclerotic Disease - Carotid Dissection, Takayasu Disease, Radiation-Induced Atherosclerosis

Spontaneous carotid dissection causes 2% of all ischemic stroke, 25% among those below 45 years of age. It is due to a tear between the intima and medial layers of the artery, creating a false lumen and intramural thrombosis. Dissection typically starts just beyond the carotid bulb, and may extend intracranially, or at least till the ICA enters the petrous temporal bone [21]. While there is still controversy as to its management by anti-platelets or anticoagulation, making a diagnosis of the cause of a patient's stroke is invaluable; recurrence is uncommon [22]. Ultrasound findings include the detection of an intimal flap (rare), or a double lumen along the ICA beyond the bulb through which blood is able to flow. Most commonly, the ICA is totally occluded just beyond the bulb.

Takayasu arteritis is an inflammatory panaortitis, more frequent in Japan, South Asia, India and Mexico, that causes cerebral ischemia by the haemodynamic effects of progressive stenosis and occlusion of the arteries. The process starts in the aorta and ascends the carotid tree. Treatment includes the use of steroids in the active stage of the disease, and bypass and reimplantation surgery in burnt-out disease. On ultrasound, the inner layers of the arterial walls is diffusely thickened, starting at the proximal CCA, and extending cephalically [23].

The risk of stroke and transient ischaemic attacks is doubled by radiotherapy to the head and neck for malignancies. Radiation-induced atherosclerosis, the thickening of the arterial wall, is probably due to accelerated atherosclerosis due to radiation to the intima-medial layer as well as injury to the vasovasorum of the adventitia [24]. Endarterectomy is difficult to perform in the scarred tissues; stenting has a high re-stenosis rate [25]. On ultrasound, the atherosclerosis is restricted to the parts of the artery that lay within the field of radiation, with a sharp cut-of in the area of abnormality between the irradiated and non-irradiated parts of the artery.

Subclavian Steal

While usually asymptomatic, the patient may complain of dizziness with excessive use of an upper limb. The pulse and blood pressure may be weaker in that limb. The underlying process is severe stenosis/occlusion of the ScA proximal to the origin of the VA. Severe symptomatic stenosis may be treated by surgery or stenting [26].

On ultrasound, the earliest indication is a transient sharp deceleration of blood flow after the first systolic peak, producing a notch and giving rise to two systolic peaks: the first is sharp, the second is blunt. As the subclavian steal worsens, the systolic notch becomes more obvious and the second systolic peak decreases. The nadir of the notch becomes lower and lower until it reaches and finally crosses the baseline. Flow reversal during systole is initially minimal, but increases until there is complete reversal of flow throughout the cardiac cycle. The steal can be accentuated by the cuff test where a blood pressure cuff is placed over the upper arm, inflated just beyond systolic blood pressure and maintained for a few minutes, and then suddenly completely released (Figure 14).

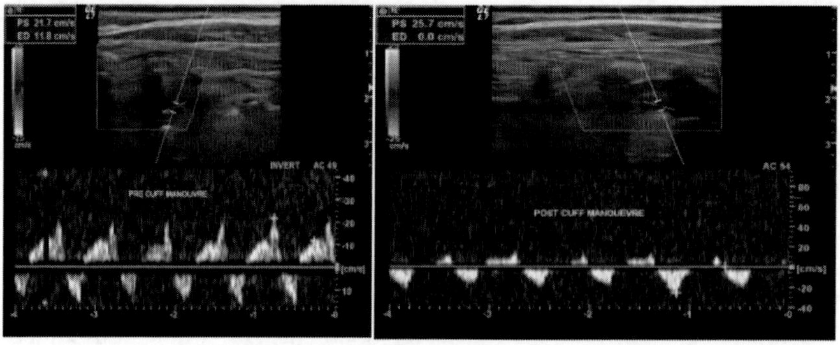

Figure 14. Subclavian steal - alternating flow in vertebral artery accentuated with cuff manoeuvre.

Intracranial Arterial Stenosis/Occlusion

Intracranial atherosclerosis accounts for about 8% of ischaemic stroke in North America and 50% in Asia. The risk of recurrent stroke is 15% per year [27]. Intracranial atherosclerosis typically occurs at sites of arterial bifurcation or merging, or areas of turbulent flow. These sites include the CS, tICA, M1, A1, P1, vertebro-basilar junction, mid-BA, and top of the BA. Optimal management is still uncertain – anti-platelets may be sufficient [28, 29].

Various ultrasound diagnostic criteria have been proposed [30]. Just as for carotid disease, each neurovascular laboratory is encouraged to develop its own validated criteria for grading stenosis.

Collateral Flow As a Result of Severe Stenosis/ Occlusion

Collateral blood flow via extracranial and intracranial anastomoses provides much needed blood supply and reduces the risk of haemodynamic stroke.

In ICA severe stenosis/occlusion between the carotid bifurcation and the origin of OA, extracranial anastomosis between the ECA and ICA via the OA results in increased ipsilateral ECA flow with reversal of the OA flow direction to enter the distal ICA.

Ipsilateral A1 flow is reversed as flow increases in the contralateral A1 and crosses the ACommA to enter the A1; ipsilateral P1 flow may also increase and across the PCommA to enter the distal ICA.

In CCA occlusion, flow may reverse along the ipsilateral ECA to feed the ipsilateral ICA. In VA occlusion, increased flow may occur in the contralateral VA to feed the BA.

Right-To-Left Shunting

Approximately 40% of ischaemic strokes among the young are cryptogenic; more than half have a patent foramen ovale that may provide a conduit for a thrombus from the right-sided circulation to cross into the left-sided circulation and enter the cerebral circulation [31]. Optimal therapy is controversial – closure, anti-coagulants, anti-platelets are used [32].

While the ‘gold standard’ for the diagnosis of PFO is transoesophageal echocardiography (TEE), TCD of the MCA has been shown to have a very high sensitivity and specificity [33]. For those patients with poor temporal windows, the foraminal window with the insonation of the BA can be used [34].

There is a consensus recommendation for the performance of a ‘bubble-TCD” for the detection of a right-to-left shunt(RLS). This involves the rapid injection via a large bore needle placed in the antecubital region of a well-agitated mixture of 9 ml of saline and 1 ml of air, while insonating the MCA.

The presence of an RLS is indicated by the swift occurrence of high intensity transient signals (HITS) presumed to be microbubbles (MB) - 0 MB (negative result); 1-10 MB; >10 MB with no curtain, and curtain [35]. The optimal position may be with the patient sitting and test performed with the Valsalva manoeuvre [36]. Approximately 10% of RLS are extra-cardiac eg pulmonary vascular malformations.

Embolisation

The presence of HITS gives a 7 to 13-fold increase in the risk of recurrent stroke [37, 38]. HITS are attributed to microemboli from a more proximal source – thrombi (eg from atrial fibrillation, prosthetic cardiac valves, atherosclerotic plaques extra or intracranially), gas bubbles (eg from prosthetic cardiac valves). They are reduced by double-antiplatelet therapy, but reduction in clinical events has not been shown [39, 40].

The TCD probe is fixed on a head frame and the artery closely monitored, preferably for at least 30 mins, on at least 3 occasions. There are consensus criteria for Doppler diagnosis – transient signal usually lasting less than 300 milliseconds, amplitude usually at least 3 dB higher than that of the background flow signal, signal is unidirectional within the Doppler spectrum, and signal is accompanied by an audible "snap," "chirp," or "moan" [41].

Intra-Operative Monitoring during Carotid Endarterectomy (CEA), Coronary Artery Bypass Grafting (CABG)

Intracranial HITS are frequently seen during CEA, especially during the mobilization of the carotid artery and during cross-clamping, and even more-so after release of the cross clamps after arterial wall closure. These are presumed to be due to emboli from the atherosclerotic plaques, or even air that entered the surgical site during surgery.

Similarly, HITS are seen during CABG. These may be due to gaseous bubbles from the heart-lung machine or atherosclerotic material during reversed flow along the aorta.

During CEA, due insufficient intracranial collaterals, MCA flow may fall precipitously during ipsilateral cross-clamping – some surgeons would insert a shunt across the surgical incision site in this setting. Monitoring is achieved by using a head frame holding the monitoring probes.

Post-Surgery/Stent Assessment

Patency of the carotid arteries immediately after CEA or stenting, intra/post-operative carotid dissection, in-situ thrombosis, are detected by ultrasound, which may require re-surgery/stent. The occurrence of subsequent re-stenosis sometime after surgery/stenting is also easily assessed. If a patient develops post-procedure ipsilateral cortical signs, the finding of extremely high ipsilateral intracranial velocities may support the diagnosis of hyper-perfusion syndrome, necessitating blood pressure reduction.

Cerebrovascular Reserve to Predict Stroke Risk

Vasodilation is one of the cerebrovascular responses to cerebral ischaemia so as to maintain blood flow when there is blood flow reduction. Carbon dioxide ($CO2$) is a potent cerebral vasodilator. A reduced or failed vasodilatory response to CO2 would indicate that vasodilation is maximal, no more vasodilation is possible ie there is absent/reduced cerebrovascular reserve (CVR). Reduced CVR is a marker for recurrent cerebral ischaemic events, and may be used to select patients for superficial temporal artery-middle cerebral artery bypass and carotid surgery [42].

The breath holding index (BHI) is one method of assessing CVR. It is determined by measuring the mean velocity before and after breath-holding: BHI = (velocity at end of breath-holding - velocity at start of breath-holding) x 100/ (velocity at start of breath-holding x number of seconds of breath-holding). A BHI < 0.69 indicates an increased risk of ipsilateral ischaemic cerebrovascular events [43].

Sonothrombolysis

Ultrasound is able to assist in the lysis of blood clots by a number of mechanisms, including clot agitation to break up clots and thus increase the surface area for natural or administered thrombolytic molecules or agents to act. The administration of ultrasound during the administration of systemic thrombolytic agents has been shown to improve recanalisation, without harm or benefit on death or dependency at 3 months [44].

The ultrasound probe is fixed to the head by a frame with the ultrasound beam focused at the site of occlusion for an hour. Recanalisation is heralded by the return of flow velocities to a more normal level at which time the insonation can be terminated.

Physician Accreditation

The performance of a competent neurosonological examination will greatly assist in patient care. There are a number of international accreditation examinations available for physicians held a number of times in the year in various cities. They are offered by the World Federation of Neurology Neurosonology Research Group as well as the American Society of Neuroimaging [45, 46].

CONCLUSION

Ultrasound is a safe, useful and versatile technique in the management of patients with cerebrovascular disease, not just for diagnosis but potentially also for treatment (sonothrombolysis).

Further research and clinical trials will provide more information as to the role of neurosonology in the management of ischaemic stroke.

REFERENCES

[1] Murray, C. J., Lopez, A. D. Alternative projections of mortality and disability by cause 1990-2020: Global Burden of Disease Study. *Lancet.* 1997 May 24;349(9064):1498-504.

[2] Feigin, V. L., Lawes, C. M., Bennett, D. A., Barker-Collo, S. L., Parag, V. Worldwide stroke incidence and early case fatality reported in 56 population-based studies: a systematic review. *Lancet Neurol.* 2009 Apr.;8(4):355-69.

[3] Hankey, G. J., Warlow, C. P. The role of imaging in the management of cerebral and ocular ischaemia. *Neuroradiology.* 1991;33(5):381-90.

[4] http://www.fda.gov/RadiationEmittingProducts/RadiationEmittingProductsandProcedures/MedicalImaging/ucm115357.htm

[5] Chambless, L. E., Heiss, G., Folsom, A. R., Rosamond, W., Szklo, M., Sharrett, A. R., et al. Association of coronary heart disease incidence with carotid arterial wall thickness and major risk factors: the Atherosclerosis Risk in Communities (ARIC) Study, 1987-1993. *Am. J. Epidemiol.* 1997 Sep. 15;146(6):483-94.

[6] O'Leary, D. H., Polak, J. F., Kronmal, R. A., Manolio, T. A., Burke, G. L., Wolfson, S. K., Jr. Carotid-artery intima and media thickness as a risk factor for myocardial infarction and stroke in older adults. Cardiovascular Health Study Collaborative Research Group. *N Engl. J. Med.* 1999 Jan. 7;340(1):14-22.

[7] Bots, M. L., Hoes, A. W., Koudstaal, P. J., Hofman, A., Grobbee, D. E. Common carotid intima-media thickness and risk of stroke and myocardial infarction: the Rotterdam Study. *Circulation.* 1997 Sep. 2;96(5):1432-7.

[8] Kitamura, A., Iso, H., Imano, H., Ohira, T., Okada, T., Sato, S., et al. Carotid intima-media thickness and plaque characteristics as a risk factor for stroke in Japanese elderly men. *Stroke*. 2004 Dec.;35(12):2788-94.

[9] Murakami, S., Otsuka, K., Hotta, N., Yamanaka, G., Kubo, Y., Matsuoka, O., et al. Common carotid intima-media thickness is predictive of all-cause and cardiovascular mortality in elderly community-dwelling people: Longitudinal Investigation for the Longevity and Aging in Hokkaido County (LILAC) study. *Biomed. Pharmacother*. 2005 Oct.;59 Suppl. 1:S49-53.

[10] Lorenz, M. W., von Kegler, S., Steinmetz, H., Markus, H. S., Sitzer, M. Carotid intima-media thickening indicates a higher vascular risk across a wide age range: prospective data from the Carotid Atherosclerosis Progression Study (CAPS). *Stroke*. 2006 Jan.;37(1):87-92.87.

[11] Crouse, J. R., 3rd, Raichlen, J. S., Riley, W. A., Evans, G. W., Palmer, M. K., O'Leary, D. H., et al. Effect of rosuvastatin on progression of carotid intima-media thickness in low-risk individuals with subclinical atherosclerosis: the METEOR Trial. *JAMA*. 2007 Mar. 28;297(12):1344-53.

[12] Roffi, M., Lüscher, T. F. Management of patients with carotid artery stenosis. *Herz*. 2008 Nov.;33(7):490-7.

[13] Rerkasem, K., Rothwell, P. M. Carotid endarterectomy for symptomatic carotid stenosis. *Cochrane Database Syst Rev*. 2011 Apr. 13;(4):CD001081.

[14] Howard, G., Baker, W. H., Chambless, L. E., Howard, V. J., Jones, A. M., Toole, J. F. An approach for the use of Doppler ultrasound as a screening tool for hemodynamically significant stenosis (despite heterogeneity of Doppler performance). A multicenter experience. Asymptomatic Carotid Atherosclerosis Study Investigators. *Stroke*. 1996 Nov.;27(11):1951-7.

[15] Von Reutern, G. M., Goertler, M. W., Bornstein, N. M., Del Sette, M., Evans, D. H., Hetzel, A., Kaps, M., Perren, F., Razumovky, A., von Reutern, M., Shiogai, T., Titianova, E., Traubner, P., Venketasubramanian, N., Wong, L. K., Yasaka, M.; Neurosonology Research Group of the World Federation of Neurology. Grading carotid stenosis using ultrasonic methods. *Stroke*. 2012 Mar.;43(3):916-21.

[16] Jahromi, A. S., Cinà, C. S., Liu, Y., Clase, C. M. Sensitivity and specificity of color duplex ultrasound measurement in the estimation of internal carotid artery stenosis: a systematic review and meta-analysis. *J. Vasc. Surg*. 2005 Jun.;41(6):962-72.

[17] Grant, E. G., Benson, C. B., Moneta, G. L., Alexandrov, A. V., Baker, J. D., Bluth, E. I., Carroll, B. A., Eliasziw, M., Gocke, J., Hertzberg, B. S., Katarick, S., Needleman, L., Pellerito, J., Polak, J. F., Rholl, K. S., Wooster, D. L., Zierler, E.; Society of Radiologists in Ultrasound. Carotid artery stenosis: grayscale and Doppler ultrasound diagnosis--Society of Radiologists in Ultrasound consensus conference. *Ultrasound Q.* 2003 Dec.;19(4):190-8.

[18] Chua, H. C., Sitoh, Y. Y., Earnest, A., Venketasubramanian, N. Detection of internal carotid artery stenosis with duplex velocity criteria using receiver operating characteristic analysis. *Ann. Acad. Med. Singapore.* 2007 Apr.;36(4):247-52.

[19] Coward, L. J., McCabe, D. J., Ederle, J., Featherstone, R. L., Clifton, A., Brown, M. M.; CAVATAS Investigators. Long-term outcome after angioplasty and stenting for symptomatic vertebral artery stenosis compared with medical treatment in the Carotid And Vertebral Artery Transluminal Angioplasty Study (CAVATAS): a randomized trial. *Stroke.* 2007 May;38(5):1526-30.

[20] Yurdakul, M., Tola, M. Doppler criteria for identifying proximal vertebral artery stenosis of 50% or more. *J. Ultrasound Med.* 2011 Feb.;30(2):163-8. Koch, S., Romano, J. G., Park, H., Amir, M., Forteza, A. M. Ultrasound velocity criteria for vertebral origin stenosis. *J. Neuroimaging.* 2009 Jul.;19(3):242-5.

[21] Fusco, M. R., Harrigan, M. R. Cerebrovascular dissections--a review part I: Spontaneous dissections. *Neurosurgery.* 2011 Jan. 68(1):242-57

[22] Kennedy, F., Lanfranconi, S., Hicks, C., Reid, J., Gompertz, P., Price, C., Kerry, S., Norris, J., Markus, H. S.; On behalf of the CADISS Investigators. Antiplatelets vs anticoagulation for dissection: CADISS nonrandomized arm and meta-analysis. *Neurology.* 2012 Aug. 14;79(7):686-689;68(1):242-57.

[23] Venketasubramanian, N. Diagnosis and management of Takayasu arteritis. *Perspectives in Medicine* 2012 (in press).

[24] Plummer, C., Henderson, R. D., O'Sullivan, J. D., Read, S. J. Ischemic stroke and transient ischemic attack after head and neck radiotherapy: a review. *Stroke.* 2011 Sep.;42(9):2410-8.

[25] Fokkema, M., den Hartog, A. G., Bots, M. L., van der Tweel, I., Moll, F. L., de Borst, G. J. Stenting versus surgery in patients with carotid stenosis after previous cervical radiation therapy: systematic review and meta-analysis. *Stroke.* 2012 Mar.;43(3):793-801.

[26] Song, L., Zhang, J., Li, J., Gu, Y., Yu, H., Chen, B., Guo, L., Wang, Z. Endovascular stenting vs. extrathoracic surgical bypass for symptomatic subclavian steal syndrome. *J. Endovasc. Ther.* 2012 Feb.;19(1):44-51

[27] Leung, T. W., Kwon, S. U., Wong, K. S. Management of patients with symptomatic intracranial atherosclerosis. *Int. J. Stroke.* 2006 Feb.;1(1):20-5.

[28] Chimowitz, M. I., Lynn, M. J., Howlett-Smith, H., Stern, B. J., Hertzberg, V. S., Frankel, M. R., Levine, S. R., Chaturvedi, S., Kasner, S. E., Benesch, C. G., Sila, C. A., Jovin, T. G., Romano, J. G.; Warfarin-Aspirin Symptomatic Intracranial Disease Trial Investigators. Comparison of warfarin and aspirin for symptomatic intracranial arterial stenosis. *N Engl. J. Med.* 2005 Mar. 31;352(13):1305-16.

[29] Chimowitz, M. I., Lynn, M. J., Derdeyn, C. P., Turan, T. N., Fiorella, D., Lane, B. F., Janis, L. S., Lutsep, H. L., Barnwell, S. L., Waters, M. F., Hoh, B. L., Hourihane, J. M., Levy, E. I., Alexandrov, A. V., Harrigan, M. R., Chiu, D., Klucznik, R. P., Clark, J. M., McDougall, C. G., Johnson, M. D., Pride, G. L. Jr, Torbey, M. T., Zaidat, O. O., Rumboldt, Z., Cloft, H. J.; SAMMPRIS Trial Investigators. Stenting versus aggressive medical therapy for intracranial arterial stenosis. *N Engl. J. Med.* 2011 Sep. 15;365(11):993-1003.

[30] Zhao, L., Barlinn, K., Sharma, V. K., Tsivgoulis, G., Cava, L. F., Vasdekis, S. N., Teoh, H. L., Triantafyllou, N., Chan, B. P., Sharma, A., Voumvourakis, K., Stamboulis, E., Saqqur, M., Harrigan, M. R., Albright, K. C., Alexandrov, A. V. Velocity criteria for intracranial stenosis revisited: an international multicenter study of transcranial Doppler and digital subtraction angiography. *Stroke.* 2011 Dec.;42(12):3429-34.

[31] 1: Meier, B., Frank, B., Wahl, A., Diener, H. C. Secondary stroke prevention: patent foramen ovale, aortic plaque, and carotid stenosis. *Eur. Heart J.* 2012 Mar.;33(6):705-13, 713a, 713b.

[32] Agarwal, S., Bajaj, N. S., Kumbhani, D. J., Tuzcu, E. M., Kapadia, S. R. Meta-analysis of transcatheter closure versus medical therapy for patent foramen ovale in prevention of recurrent neurological events after presumed paradoxical embolism. *JACC Cardiovasc. Interv.* 2012 Jul.;5(7):777-89.

[33] Di Tullio, M., Sacco, R. L., Venketasubramanian, N., Sherman, D., Mohr, J. P., Homma, S. Comparison of diagnostic techniques for the detection of a patent foramen ovale in stroke patients. *Stroke.* 1993 Jul.;24(7):1020-4.

[34] Venketasubramanian, N., Sacco, R. L., Di Tullio, M., Sherman, D., Homma, S., Mohr, J. P. Vascular distribution of paradoxical emboli by transcranial Doppler. *Neurology*. 1993 Aug.;43(8):1533-5.

[35] Jauss, M., Zanette, E. Detection of right-to-left shunt with ultrasound contrast agent and transcranial Doppler sonography. *Cerebrovasc. Dis.* 2000 Nov.-Dec.;10(6):490-6.

[36] Agustin, S. J., Yumul, M. P., Kalaw, A. J., Teo, B. C., Eng, J., Phua, Z., Singh, R., Gan, R. N., Venketasubramanian, N. Effects of posture on right-to-left shunt detection by contrast transcranial doppler. *Stroke*. 2011 Aug.;42(8):2201-5.

[37] Ritter, M. A., Dittrich, R., Thoenissen, N., Ringelstein, E. B., Nabavi, D. G. Prevalence and prognostic impact of microembolic signals in arterial sources of embolism. A systematic review of the literature. *J. Neurol.* 2008 Jul.;255(7):953-61.

[38] King, A., Markus, H. S. Doppler embolic signals in cerebrovascular disease and prediction of stroke risk: a systematic review and meta-analysis. *Stroke*. 2009 Dec.;40(12):3711-7.

[39] Markus, H. S., Droste, D. W., Kaps, M., Larrue, V., Lees, K. R., Siebler, M., Ringelstein, E. B. Dual antiplatelet therapy with clopidogrel and aspirin in symptomatic carotid stenosis evaluated using doppler embolic signal detection: the Clopidogrel and Aspirin for Reduction of Emboli in Symptomatic Carotid Stenosis (CARESS) trial. *Circulation*. 2005 May 3;111(17):2233-40 40.

[40] Wong, K. S., Chen, C., Fu, J., Chang, H. M., Suwanwela, N. C., Huang, Y. N., Han, Z., Tan, K. S., Ratanakorn, D., Chollate, P., Zhao, Y., Koh, A., Hao, Q., Markus, H. S.; CLAIR study investigators. Clopidogrel plus aspirin versus aspirin alone for reducing embolisation in patients with acute symptomatic cerebral or carotid artery stenosis (CLAIR study): a randomised, open-label, blinded-endpoint trial. *Lancet Neurol.* 2010 May;9(5):489-97.

[41] Basic identification criteria of Doppler microembolic signals. Consensus Committee of the Ninth International Cerebral Hemodynamic Symposium. *Stroke*. 1995 Jun.;26(6):1123.

[42] Widder, B., Paulat, K., Hackspacher, J., Mayr, E. Transcranial Doppler CO2 test for the detection of hemodynamically critical carotid artery stenoses and occlusions. *Eur. Arch. Psychiatry Neurol. Sci.* 1986;236(3):162-8.

[43] Vernieri, F., Pasqualetti, P., Passarelli, F., Rossini, P. M., Silvestrini, M. Outcome of carotid artery occlusion is predicted by cerebrovascular reactivity. *Stroke*. 1999 Mar.;30(3):593-8.

[44] Ricci, S., Dinia, L., Del Sette, M., Anzola, P., Mazzoli, T., Cenciarelli, S., Gandolfo, C. Sonothrombolysis for acute ischaemic stroke. *Cochrane Database Syst. Rev*. 2012 Jun. 13;6:CD008348.

[45] http://nsrg.net/education/certification/item/48-guidelines.html

[46] http://www.asnweb.org/i4a/pages/index.cfm?pageid=3604.

In: Ultrasonography
Editor: Alexander Burdenko

ISBN: 978-1-62417-536-7

Chapter 3

ROLE OF ENDOSCOPIC ULTRASONOGRAPHY IN PANCREATIC SOLID MASSES

Zaida Adrián-de Ganzo*, Antonio Z. Gimeno-García*,† and Enrique Quintero

Servicio de Aparato Digestivo, Hospital Universitario de Canarias, La Laguna, Tenerife, España

ABSTRACT

Endoscopic ultrasonography is a combination of endoscopic view and ultrasonography. It was developed to improve diagnostic accuracy by directly imaging the target lesion via the gastrointestinal lumen. The examination of the pancreatic gland is one of the most common indications of the endoscopic ultrasonography because of the vicinity of the endoscopic ultrasonography probe located in the stomach or duodenum to the pancreas. In fact, endoscopic ultrasonography has been considered the most accurate imaging technique in terms of detection rate and locoregional staging, particularly in tumors smaller than 2 cm in size. This technique may also be useful for clarifying abnormalities detected by other techniques such as the computed tomography or the magnetic resonance imaging. Diagnosis of solid lesions located in the pancreatic

---

* Zaida Adrián-de Ganzo and Antonio Z Gimeno-Garcia contributed equally to this work.
† Corresponding address: Antonio Z Gimeno-García, MD, PhD. Hospital Universitario de Canarias. Unidad de Endoscopia (10ª planta). Departamento de Gastroenterología. Ofra s/n.38320.La Laguna. Tenerife. Spain.

gland are usually challenging for the clinician. Although most of them are adenocarcinomas, diagnosis is not always straightforward and a cytological or histological sample is sometimes mandatory because clinical management may be different depending on the diagnosis. Unusual solid lesions such as pancreatic metastasis, pancreatic lymphoma or neuroendocrine tumors account for 10% of the pancreatic masses and should be suspected based on patient´s age, family history of pancreatic cancer, past personal history of cancer and location of the tumor. Other conditions such as focal chronic pancreatitis or autoinmune pancreatitis may mimic a pancreatic adenocarcinoma and in these cases fine needle aspiration biopsy is required to rule out cancer. A cytological diagnosis is mandatory as a proof of malignancy in non-resectable pancreatic tumors before initiation of chemotherapy. In this way, endoscopic ultrasonography has advantages over other imaging techniques when we are dealing with solid pancreatic lesions including real time imaging and a lower risk of complications when an endoscopic ultrasound -guided fine needle aspiration cytology is required. In this chapter we review the indications of endoscopic ultrasonography and endoscopic ultrasound - guided fine needle aspiration cytology in solid pancreatic masses including diagnosis, indications of cytology and its usefulness in staging.

1. INTRODUCTION

Epithelial tumors, are the most frequent pancreatic malignancies and the majority originate from the exocrine portion of the gland [1]. Among them, pancreatic adenocarcinoma (PA) is the leading histological type accounting for 90% of the solid pancreatic neoplasms. This tumor ranks as one of the most lethal malignancies in the world accounting for 266669 deaths in 2008 [2]. The differential diagnosis of the remaining solid pancreatic tumors includes an heterogeneous group of lesions such as, islet cell tumors, pancreatic metastasis, solid pseudopapillary tumors, lymphomas, giant dell tumors, mesenchimal tumors and lipomas [3-5]. Despite those tumors are less common, they are becoming more frequently diagnosed because of the technological improvement and the increased utilization of the diagnostic imaging. In addition, to achieve a correct diagnosis of those tumors is crucial, as the prognosis and the management are completely different to that of the much more frequent pancreatic adenocarcinoma. Endoscopic ultrasonography (EUS) combines the endoscopic view and the ultrasound imaging. With several technological improvements, such as the introduction of electronic transducers and color-Doppler capability, EUS is able to gain accurate images.

Due to the small distance between the probe and the pancreas and the capability of tissue acquisition, EUS is considered the best imaging technique for the study of pancreas. This technique has the ability to visualize the details of the pancreatic parenchyma, the pancreatic duct, and the surrounding structures without being hindered by the presence of gas or obesity. EUS diagnostic accuracy for detecting pancreatic tumors ranged from 85 to 100%, across the studies. The high sensitivity is held even in small solid pancreatic tumors (< 2 cm ranging from 83% to 100% [6-8]. Interestingly, some authors have documented a negative predictive value for tumor detection as high as 100% [9]. Several studies comparing EUS and computed tomography (CT) confirmed the superiority of EUS in this setting.[6-8] Additional benefits include real-time imaging and low risks because of the proximity of the needle to the lesion (bleeding perforation or tumor dissemination in the needle tract) [10]. Another advantage over other imaging techniques is that EUS-guided fine needle aspiration cytology (EUS-FNA) or biopsy (EUS.FNB) may provide good quality cytological and histological specimens of pancreatic tumors [11, 12]. The ability to provide core samples or adequate material for cell block is mandatory to perform ancillary techniques such as inmunohistochemistry, flow cytometry, or karyotyping. These techniques are extremely important for the diagnosis of neuroendocrine tumors, pancreatic metastasis or sub-typing lymphoma. This chapter will highlight the usefulness of EUS and EUS-FNA in the diagnosis of solid pancreatic tumors.

2. Importance of a Correct Diagnosis

Although most of the solid pancreatic tumors are adenocarcinomas, the diagnosis is not always straightforward because in 10% of the cases less common pancreatic lesions such as islet cell tumors, pancreatic metastasis, lymphomas and other infrequent masses can be found [3-5]. The correct diagnosis of the pancreatic lesions is crucial as the prognosis and management of those tumors are completely different. Factors such as the patient´s age, family history of pancreatic cancer, past personal history of cancer, location of the tumor, as well as atypical clinical or imaging findings are key elements to establish the diagnostic suspicion [13]. For instance, a pancreatic mass in a young patient may raise the suspicion of a neuroendocrine tumor, a lymphoma or a solid pseudopapillary tumor specially in women, and, a past history of cancer may be a hint of a pancreatic metastasis. EUS imaging may also suggest one type of pancreatic tumor, however, it is not specific.

3. Role of EUS-FNA in Pancreatic Masses

As previously mentioned, EUS-FNA may be performed for cytological tissue sampling of suspicious lesions and regional lymph nodes. Its feasibility varies from 90% to 98% whereas sensitivity and specificity in series of non-selected solid pancreatic tumors ranged from 60% to 94%[14,15] and 88%-100% [16,17] respectively with an accuracy between 72% and 96%[18,19] across the studies. However, the sensitivity and specificity rate can be surprisingly variable due to a difficult cytological interpretation of low cellular samples, in well differentiated tumors, extensive fibrosis (desmoplasia) and the presence of calcifications that can prevent an adequate tissue sampling and analysis [31]. Despite these limitations, EUS-FNA can still yield better sensitivity and specificity than ERCP brushing and CT/US-guided FNA in the diagnosis of pancreatic cancer [32, 33]. Moreover, EUS-FNA can visualize and sample small lesions as neuroendocrine tumors or pancreatic metastasis, difficult to reach with a percutaneous access [34]. Overall diagnostic yield as well as sensitivity, specificity and accuracy, seem not to be influenced by tumor size [35].

However, some clinical considerations should be taken into account before taking the decision of EUS-FNA. Whether or not the biopsy will affect the management and if benefits outweigh potential risks (i.e. risk of seeding or haemorrhage) are two important questions. Patients with non-resectable pancreatic tumors will need cytological diagnosis before initiating of chemotherapy. In advanced tumors (i.e. with liver metastasis, mediastinal lymph nodes or suspicion of peritoneal carcinomatosis) it is useful to undergo EUS-FNA as a proof of malignancy [20]. Although percutaneous biopsy is an option in these patients, diagnostic accuracy is probably lower and needle tract metastasis is more frequent, painful and difficult to treat [21]. EUS-FNA has lower risk of seeding than the percutaneous approach due to a short needle track and does not pass through peritoneal or pleural surfaces. For instance, a higher frequency of peritoneal carcinomatosis after chemotherapy has been described in patients who underwent a pancreatic CT biopsy compared with those who had EUS-FNA [22].

However, even using EUS-FNA, some controversy may persist if preoperative tissue diagnosis is needed, in the case of resectable disease. Seeding is not of great importance in head or uncinate lesions, since the biopsy tract would be removed as part of the surgical specimen. In general, it should be avoided in resectable body and tail pancreatic masses, since the biopsy tract will not be included in the surgical specimen and there is a risk of needle tract seeding [23]. However, in some cases, EUS-FNA should be considered in order to exclude other pathology (i.e., lymphoma, small cell carcinoma, etc.) and to optimize surgical management (organ-preserve surgery). Another important indication is in the setting of patients with indeterminate lesions by CT/EUS (i.e. focal pancreatitis, biliary obstruction but without obvious mass, and atypical clinical presentation). EUS-guided core needle biopsy (EUS-CNB) is another tissue sampling method which provides and histological sample. It was developed to overcome the limitations of EUS-FNA, especially the low cellular sample, when a core sample is required for diagnosis. Large needles (19 gauge) have been frequently used in lesions in which additional techniques other than cytological analysis are necessary for achieving a definitive diagnosis. They seem to be principally indicated in those endoscopy units in which immediate cytological evaluation is not available [24].

4. Endoscopic Ultrasound in Pancreatic Cancer Staging

The staging of pancreatic cancer is based on the TNM classification published by American Joint Commitee on Cancer (AJCC) for pancreatic adenocarcinoma,[25] which describes the tumor extension (T), lymph node involvement (N), and distant metastases (M) (Table 1).

Table 1. TNM classification, American Joint Commitee on Cancer (AJCC) for pancreatic adenocarcinoma [25]

T stage		N stage		M stage	
Tis	Carcinoma in situ	N0	No lymph nodes	M0	No metastasis
T1	Tumor limited to the páncreas 2 cm or less in size	N1	Regional lymph node metastasis	M1	Distant metastasis
T2	Tumor limited to the páncreas more than 2 cm in size				
T3	Tumor extention beyond the pancreas				
T4	Tumor involves celiac axis and /or superior mesenteric artery (unresectable)				

4.1. T Staging and Vascular Invasion

In this classification, it is important the distinction between T3 and T4 as it reflects the resectability of the lesion. Currently, limited venous invasion does not represent an absolute contraindication for surgery. In fact, tumor involvement of the superior mesenteric or portal veins will usually be classified as T3 whereas involvement of the celiac axis, hepatic artery or superior mesenteric artery will be considered advanced local invasion and preclude surgical resection (T4) [25].

Relevant vascular structures may be readily assessed by EUS. Several predictors of invasion by EUS have been reported including irregular vascular wall, intravascular filling defects, loss of hyperecogenic interface and proximity of the tumor (less than 3 mm) to the vessel (Figure 2) [26-28]. Signs like significant collateral formation and complete vascular encasement provide a proof of venous vascular invasion with a specificity of 94% (Figure 3) [26]. Overall, the reported accuracies for T staging by EUS in pancreatic cancer ranged from 62 to 94% across the studies [26, 29].

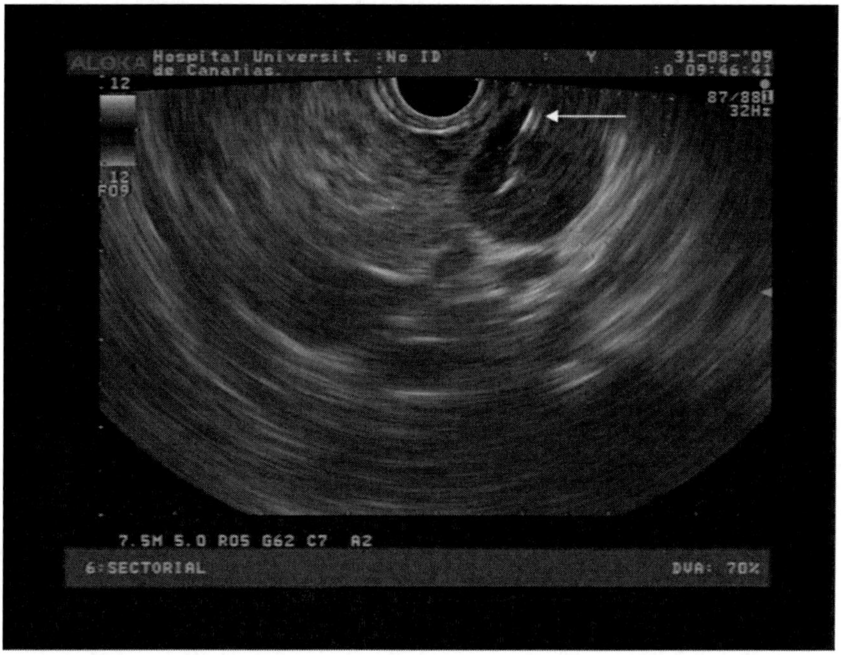

Figure 1. Endoscopic ultrasonography guided fine needle aspiration cytology (arrow) of a malignant lymph node.

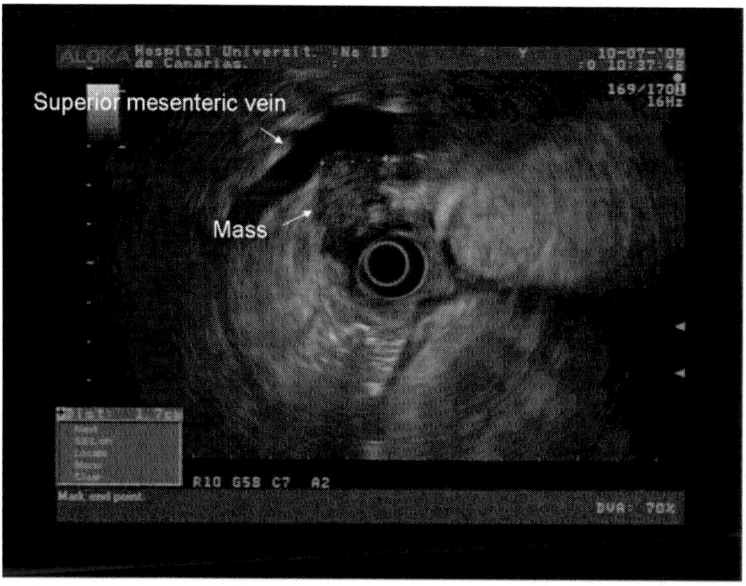

Figure 2. Pancreatic mass with vascular invasion. Note the lack of the interface between the mass and the superior mesenteric vein.

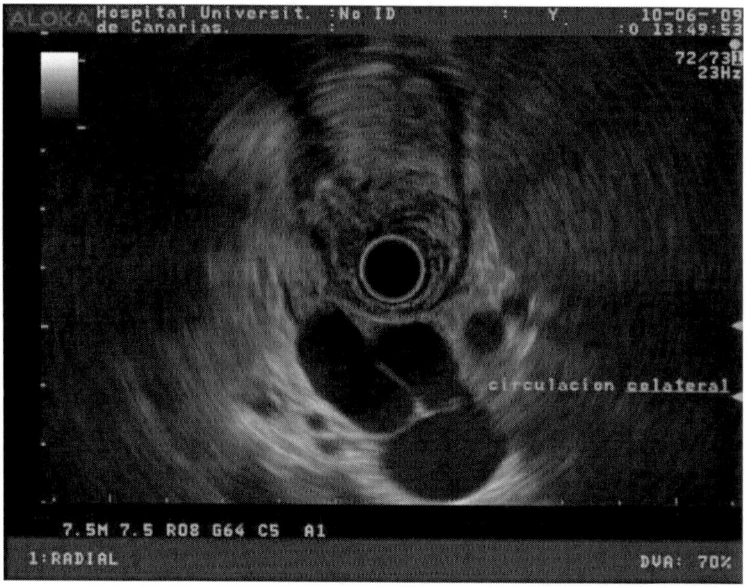

Figure 3. Collateral circulation. Note that the vessels do not go though the wall, ruling out the diagnosis of varices.

For the overall evaluation of vascular invasion the accuracy of EUS ranged from 40 to 100% with a sensitivity and specificity ranging from 42% to 91% and from 89% to 100%, respectively [30, 31]. A recent review compared the role of EUS and CT in staging of the pancreatic cancer. EUS was superior to CT for T staging and for detecting portal vein invasion while the invasion of celiac and superior mesenteric arteries were better detected by CT.[32] Overall, the prediction of resectability did not differ between both techniques.

4.2. Nodal Staging

The main node stations to be evaluated in pancreatic cancer are the perigastric, periduodenal, and celiac nodes, as well as the hepatic helium [26]. Although EUS is highly sensitive for detecting regional lymph nodes, it has difficulties to distinguish between malignant and inflammatory lymph nodes. Various criteria have been proposed for endosonographic detection of metastatic lymph nodes, including size greater than 1 cm, hypoechoic echogenicity, sharp borders, and round shape (Figure 4) [33].

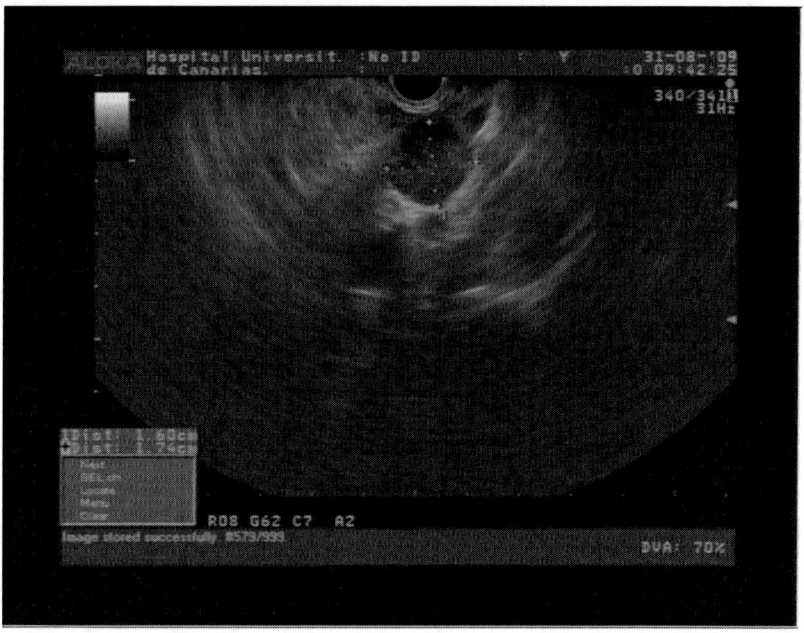

Figure 4. Celiac lymph node with features of malignancy: round shape, well demarked. hypoechoigenic, and larger than 1 cm in size.

When all four features are present within a lymph node, there is from 80% to 100% chance of malignancy. However, only about 25%of lymph nodes will have all four features present at once [34]. The sensitivity and specificity of EUS alone for the diagnosis of metastatic adenopathy in pancreatic cancer ranged from 28% to 92%[33,35] and from 26% to 100% [31,33] respectively. The accuracy of EUS for N staging of pancreatic tumors ranged from 41% to 86% [7, 36]. In some cases, when it has an impact in clinical management, the performance of a EUS-guided fine needle aspiration (FNA) of lymph nodes is necessary.

5. Pancreatic Adenocarcinoma

Pancretic adenocarcinoma is considered to have one of the worst prognoses of all solid malignancies. It has a relatively high incidence as well, making it one of the top ten incidence cancer in Europe and USA, with an overall 5-year survival rate of less than 5% [37, 38].

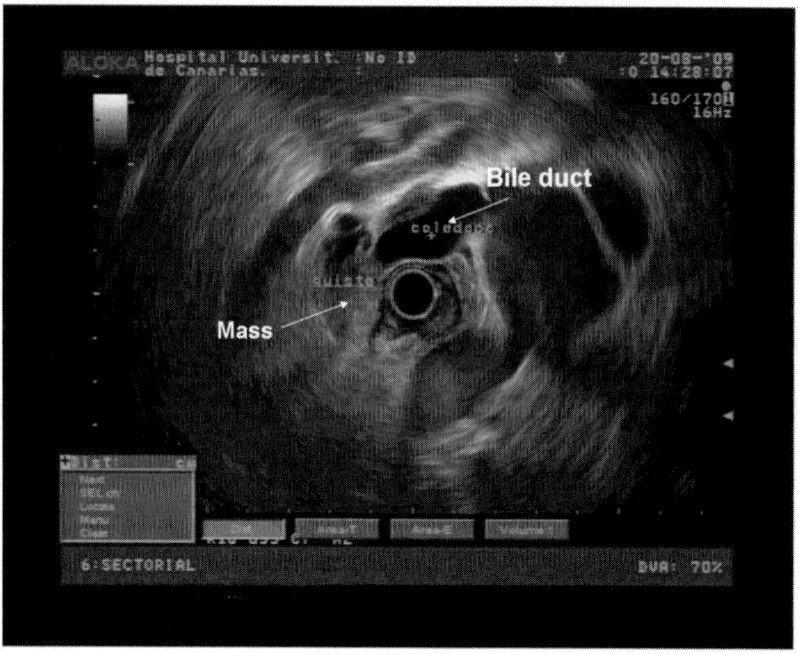

Figure 5. Malignant obstruction of the bile duct showing upstream dilation.

Several studies have shown that patients with early stage pancreatic lesions (<3 cm) without lymphatic invasion have a significantly better prognosis with a 5-year survival rate of up to 25–30% following surgical resection. These results support that early tumor detection is still essential to improve patient survival [39, 40]. However, these tumors have a non-specific clinical presentation, mainly depending on the location and the stage of the tumor and screening is not indicated in average risk population due to the low incidence of this tumor. PA has a different ultrasound appearance, depending both on tumor location and on the lesion intrinsic nature (*i.e.*, cystic component). By EUS, it is usually seen as an ill-defined, hypoechoic mass (Figure 2) [41].

It may have inhomogeneous pattern and can cause dilation of the distal pancreatic and bile duct (Figure 5) [41]. Large tumors often show areas of central citification due to tumor liquefaction. Sensitivity of EUS in pancreatic mass detection can reach 97%, with a better yield than transabdominal US and conventional CT scan, mainly with small size lesions (<2 cm) [41]. Despite high sensitivity, specificity is quite limited, especially in the differential diagnosis with inflammatory processes (i.e. focal chronic pancreatitis), which can mimic the same sonographic morphology [42].

6. Pancreatic Neuroendocrine Tumors

Pancreatic neuroendocrine tumors (PNET) arise from the endocrine pancreatic tissue. They account for 1-5% of all the pancreatic tumors and include functional and non-functional tumors [43]. Functional tumors have a characteristic clinical picture depending on the hormonal activity (table 2). Although non-functional PNET may produce hormones, they are secreted in small amounts or they are clinically inactive. Whereas functional PNET are usually diagnosed when less than 2 cm in size (Figure 6), non-functional are usually larger ranging from 2 cm to 24 cm with symptoms related to the tumor size rather than to hormone secretion [44]. Because of the improvement in imaging quality, nowadays, PNET are more frequently diagnosed even at early stages. In fact, in recent years, the percentage of non-functional PNET diagnosed incidentally have been more frequently documented involving approximately 15%-30% of PNET [45]. The most frequent PNET is the insulinoma accounting for 60% of the cases. This tumor is usually benign (90-95%), isolated, small (< 2 cm) and located in the body or tail although it can be found in anywhere [43].

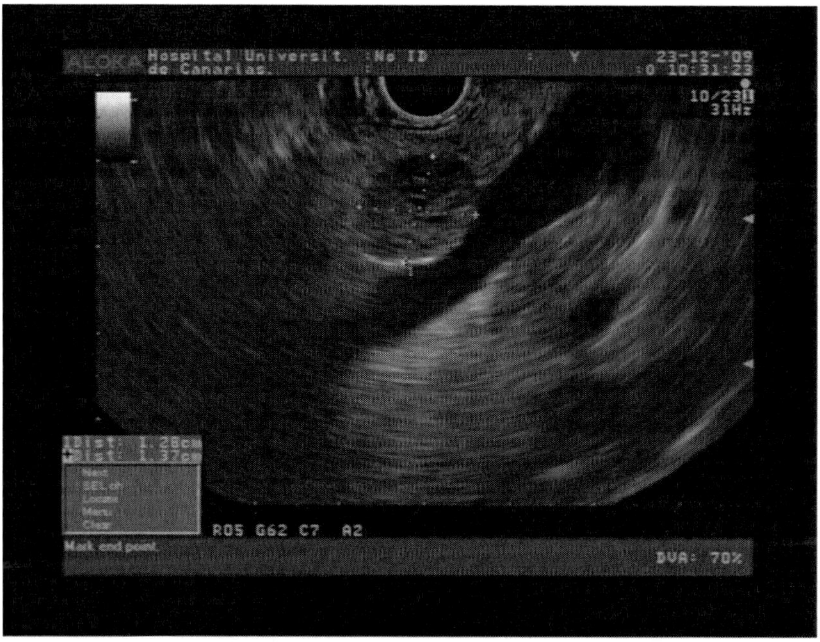

Figure 6. Neuroendocrine tumor located in the pancreatic body. Note the hipoechogenic appearance and regular margins.

Gastrinoma is the second in frequency (20%). Unlike insulinomas, they are usually malignant even when small (< 1 cm) and may lead to regional lymph node or liver metastasis. They are more frequently found in the pancreatic body or neck and sometimes they are multiple and associated with medical syndromes such as multiple endocrine neoplasia type I. Other PNETs such as glucagonomas, VIPomas and somatostatinomas are usually larger and much less frequent representing up to 10%. Diagnosis of the functional PNET is usually based on a characteristic clinical picture and clinical evidence of hormonal over-expression [43]. However, tumor location is mandatory in order to guide therapy management. In fact, for the sporadic PNET organ-preserving resection is recommended ranging from enucleation (specially in insulinomas), resection of the tail or central segment of the pancreas or pylorus-preserving resection of the pancreatic head or pancreaticoduodenal resection [43]. Cross sectional imaging studies show a sensitivity for localizing the primary tumor of 10–50%. Most PNET have high density rate of somatostatin receptors and hence, they may be detected by somatostatin receptor scintigraphy in 30%–70% of the cases [46]. The exception is the insulinoma, frequently missed by this technique (> 50%) [47].

Table 2. Classification of endocrine pancreatic tumors

Tumor	Hormone	Malignancy	Symptoms	Location
Insulinoma	Insulin proinsulin	10%	Hypoglycemia Neunoglycopenic symptoms	Anywhere (more frequent in body and tail)
Gastrinoma	Gastrin	60%-90%	Peptic ulcer disease, diarrhea	Anywhere (more frequent in body and neck)
Glucagonoma	Glucagon	50%-90%	Necrolytic migratory erythema, diabetes, depression, deep venour thrombosis	Almost excludively within body and tail
VIPoma	Vasoactive intestinal peptide	40%-70%	Secretory diarrhea, hypokalemia, achlorhydria	Anywhere
Somatostatinoma	Somatostatin	70%-80%	Diabetes, cholelithiasis, steatorrhea	Anywhere
Non-functional tumors	Pancreatic polypeptide	60%-80%	Abdominal pain, weight loss, jaundice	Anywhere

This drawback may be overcome by EUS. In addition, this technique is more precise for location and it allows tissue sampling. By EUS PNETs tend to be hypoechoic, homogeneous, with well demarcated, regular margins, and sometimes display posterior echo-enhancement owing to their vascularity. In the detection of such tumors, EUS is highly accurate and cost effective [6, 48]. Although EUS-FNA provides tissue confirmation of suspected primary or metastatic PNETs, it is not always needed, at least for functional PNETs. Cytology usually shows a monotonous, poorly cohesive population of small cells with plasmacytoid morphology. Inmunohistochemistry plays an important role, confirming the neuroendocrine nature of tumor cells.

Neuron-specific enolase, chromogranin A, synaptophysin, and neural cell adhesion molecule are usually positive in PNETs with a sensitivity and specificity exceeding 90% [48].

7. Pancreatic Metastases

Metastatic lesions of the pancreas are rare and account for about 2% of all pancreatic malignancies [49-51]. Most patients have widespread systemic diseases at the time of diagnosis and the detection of a solitary, resectable pancreatic metastasis is uncommon. The most common primary cancers associated with pancreatic metastasis are lung, renal cell, breast, colon, melanoma, and soft tissue tumors [52].

Metastasis to the pancreas may occur many years after diagnosis of the primary tumor, and this is especially true for the renal cell cancer with a median time period between nephrectomy and the finding of pancreatic metastasis of more than 10 years [53]. A recent meta-analysis including 220 patients with pancreatic metastasis, revealed that the primary tumor site was most commonly the kidney in 70.5% [54]. Obtaining a detailed medical history of previous malignancy may raise the suspicion of this diagnosis. EUS features of pancreatic metastases appear to be different from those found in the pancreatic adenocarcinoma. The appearance of metastases on ultrasound is usually homogeneous, round and well-circumscribed lesions [4, 55]. It is important to keep in mind that multiple pancreatic metastases may be present and therefore a careful examination of the whole gland should be done. Accurate diagnosis is mandatory when suspected and requires tissue sampling (cytology and cell block analysis) as well as subsequent immuno-histochemistry analysis [56].

8. Pancreatic Lymphoma

Primary pancreatic lymphoma (PPL) is extremely rare accounting for less than 0.5% of the pancreatic tumors [3]. Although, almost 30% of the non-Hodgkin lymphomas involves the pancreatic gland, less than 1% are considered primary pancreatic lymphoma (PPL) [57]. In one study it was shown to comprise 1.3% of 1050 pancreatic fine needle aspirations [58].

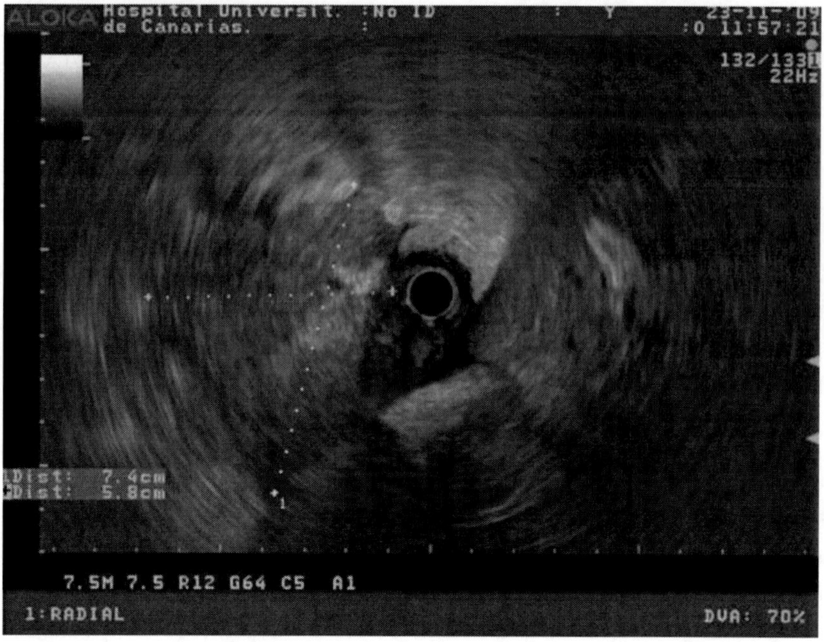

Figure 7. Pancreatic lymphoma. It is characteristic the irregular margins and the strongly hypoechogenic appearance.

The diagnostic criteria for PPL include [59]: 1) mass which predominantly affects the pancreas, 2) peripancreatic node involvement, 3) absence of palpable lymph nodes, 4) absence of mediastinal involvement, 5) absence of hepatosplenic dissemination, 6) blood cell differential count normal. The head of the pancreas is the most common location for these tumors but the entire gland may be affected.

The majority of PPL displays a B-cell phenotype. EUS pattern has been described as a "strongly hypoecogenic appearance in the pancreas, hypertrophy in all its segments, a hyperechoic wall in the common pancreatic duct and multiple isoechogenic pancreatic nodes" (Figure 7) [3, 60]. However, this pattern is not specific enough to differentiate PPL from other solid pancreatic masses.

When suspected, a cytological or histological sample is needed in order to prescribe a proper treatment [61]. Ancillary techniques such as flow cytometry and karyotype are important to achieve an accurate diagnostic tool which then obviates the need for invasive surgery to only those patients with non-diagnostic FNAs [62, 63].

9. SOLID PSEUDOPAPILLARY TUMOR

These rare pancreatic neoplasms tend to affect young women and have characteristic histological features, including a combination of solid and cystic components and the presence of pseudopapillae. Interestingly, unlike ductal adenocarcinoma, solid pseudopapillary tumors of the pancreas (SPTs) are not associated with abnormalities in K-*ras*, p53, or DPC4 genes but like colon cancer, do demonstrate genetic abnormalities in the APC/*B*-catenin *pathway [64].*

Symptoms are frequently non-specific in nature and often include abdominal pain, increased abdominal girth, and the presence of a palpable mass. A review of 718 patients with solid pseudopapillary tumors of the pancreas found a 10:1 ratio of affected women to men, a mean age of 22 years, the presence of metastases in 20% of the patients, and a cure rate greater than 95% [65]. Although the majority of these tumors are considered to be of low malignant potential, a recent report of two cases of aggressive SPTs showed death of the patients at 6 and 16 months after diagnosis.

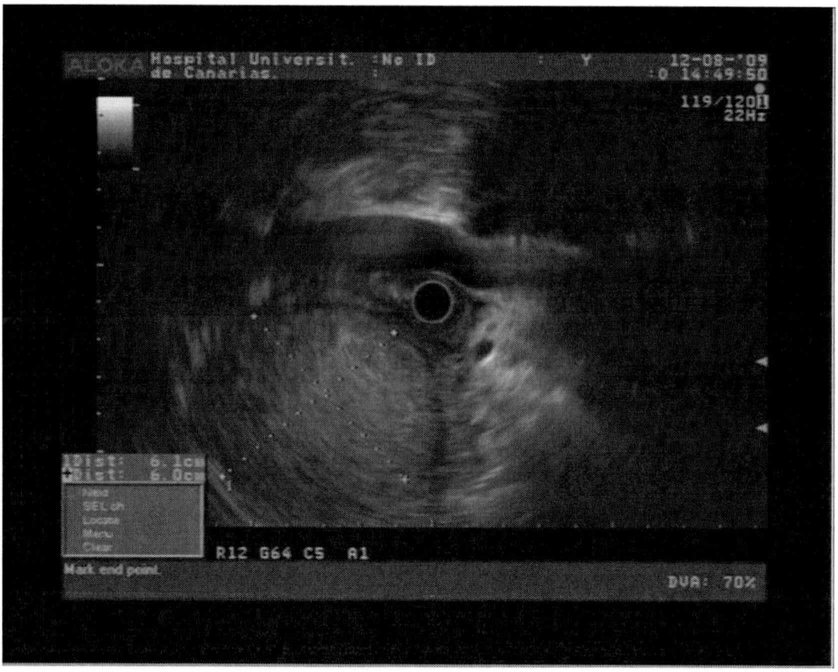

Figure 8. Pseudopapillary tumor. Note the solid and encapsulated appearance.

In these cases a high mitotic index and extensive tumor necrosis was noted [66]. Surgery is the treatment of choice for these tumors, with pancreaticoduodenectomy performed for lesions in the head of the pancreas and distal pancreatectomy for those lesions in the body or tail. Metastatic or locally invasive disease is not necessarily a contraindication to surgery, because improved long term survival has been demonstrated with aggressive surgical resection [67]. There have been reports of some success with radiotherapy or chemotherapy in cases deemed unresectable [65]. SPT is often suspected during complementary imaging investigations such as ultrasound or CT scan of the abdomen, usually showing a well-encapsulated complex mass with both solid and cystic components. A recent EUS multicenter study [68]. including 28 pseudopapillary tumors depicted them as echo–poor lesions with well–defined margins, half of them solid whereas 39% were solid and cystic and only 11% cystic (Figure 8). Ocassionally, central or periferic calcifications were found. A preoperative diagnosis was made in most lesions (75%).

10. Giant Cell Tumors

Giant cell tumors (GCT) are rare pancreatic tumors, accounting for less than 1% of all pancreatic malignancies [69]. Three types of giant cell tumors are described: osteoclastic, pleomorphic, and mixed [70]. Osteoclastic giant cell tumor (OGCT) resembles giant cell tumor of the bone, containing benign appearing osteoclastlike multinucleated cells and mononuclear cells. Pleomorphic giant cell tumor (PGCT) of the pancreas is a highly anaplastic neoplasm with bizarre pleomorphic mononucleated and multinucleated giant cells. The mixed variety has features seen in both OGCT and PGCT. Similar to pancreatic adenocarcinoma, giant cell tumors of the pancreas tend to occur in the elderly and either in women or in men [71, 72]. The most frequent symptom is the epigastric pain and jaundice due to malignant biliary obstruction. These tumors are malignant and prognosis is similar to that of the pancreatic adenocarcinoma [69]. EUS appearance of GCT differs from typical pancreatic adenocarcinoma [72]. In fact, these tumors are usually larger, they tend to be more heterogeneous with well demarcated hyperechoic and hypoechoic areas closely opposed within the same lesion. Despite their large size, vessels usually remain spare. However, these echographic features are not patognomonic of these tumors and therefore, EUS-FNA is mandatory. In a recent study, the diagnosis of GCT, including the subtype, was made by EUS-FNA in all the patients included [72].

11. Mesenchymal Tumors

Mesenchymally derived tumors of the pancreas are exceedingly rare [73, 74]. The pancreatic mesenchyme encompasses the unspecialized cells that produce the connective tissue, neuronal, vascular, and lymphatic elements. Gastrointestinal stromal tumors (GIST) are particularly interesting in this group as they are potentially malignant [75].

In the pancreas they have been depicted as hypoechoic and heterogenic masses. Similarly to GIST located in other sites, immunohistochemical analysis confirming the positivity for CD-117 is the gold standard for diagnosis. Schwannomas of the pancreas are usually benign although those associated with von Recklinghausen's (NF2) disease may be malignant and can carry a worse prognosis [74, 76].

12. Pancreatic Lipoma

Fewer than 15 cases of pancreatic lipoma have been described in the literature. Only two cases have been reported using EUS-FNA [77, 78]. Characteristics are not different from other locations being usually described as an well-circumscribed hyperecogenic mass. Although FNA is not mandatory when these tumors are suspected, lyposarcomas of the pancreas have been described.

Cytologic examination shows mature adipocytes characterized by a large fat droplet and eccentrically placed small nuclei accompanied by thin-walled vessels and no atypical cells.

Conclusion

Although pancreatic adenocarcinoma is by far the most frequent tumor in the pancreas, awareness of the other solid lesions is warranted as clinical management differs from that of the pancreatic adenocarcinoma. EUS and EUS-FNA are complementary tools that provide an accurate diagnosis of solid pancreatic lesions with low risk of complications and achieving high diagnostic rates. In addition EUS is a complementary imaging technique for the loco-regional staging of solid pancreatic neoplasms with an important role either in the vascular or in the nodal staging.

REFERENCES

[1] Nakamura E, Shimizu M, Itoh T, Manabe T. Secondary tumors of the pancreas: clinicopathological study of 103 autopsy cases of Japanese patients. *Pathology international* 2001; 51:686-690 2.http://globocan.iarc.fr/factsheets/populations/factsheet.asp?uno=900.

[2] Gimeno-Garcia AZ, Alonso MM, Garcia Castro C, Nicolas Perez D, Quintero E. [Primary pancreatic lymphoma diagnosed by endoscopic ultrasound-guided fine needle aspiration biopsy]. *Gastroenterologia y hepatologia*; 33:638-642.

[3] Gimeno-Garcia AZ, Fernandez-Esparrach G, Sendino O, Gines A. [Diagnosis of pancreatic metastasis through endoscopic ultrasonography-guided aspiration puncture]. *Medicina clinica* 2007; 129:119.

[4] Gines A, Vazquez-Sequeiros E, Soria MT, Clain JE, Wiersema MJ. Usefulness of EUS-guided fine needle aspiration (EUS-FNA) in the diagnosis of functioning neuroendocrine tumors. *Gastrointestinal endoscopy* 2002; 56:291-296.

[5] Ardengh JC, Rosenbaum P, Ganc AJ, et al. Role of EUS in the preoperative localization of insulinomas compared with spiral CT. *Gastrointestinal endoscopy* 2000; 51:552-555.

[6] DeWitt J, Devereaux B, Chriswell M, et al. Comparison of endoscopic ultrasonography and multidetector computed tomography for detecting and staging pancreatic cancer. *Annals of internal medicine* 2004; 141:753-763.

[7] Yasuda K, Mukai H, Cho E, Nakajima M, Kawai K. The use of endoscopic ultrasonography in the diagnosis and staging of carcinoma of the papilla of Vater. *Endoscopy* 1988; 20 Suppl 1:218-222.

[8] Klapman JB, Chang KJ, Lee JG, Nguyen P. Negative predictive value of endoscopic ultrasound in a large series of patients with a clinical suspicion of pancreatic cancer. *The American journal of gastroenterology* 2005; 100:2658-2661.

[9] Paquin SC, Gariepy G, Lepanto L, et al. A first report of tumor seeding because of EUS-guided FNA of a pancreatic adenocarcinoma. *Gastrointestinal endoscopy* 2005;61:610-611.

[10] Crowe DR, Eloubeidi MA, Chhieng DC, et al. Fine-needle aspiration biopsy of hepatic lesions: computerized tomographic-guided versus endoscopic ultrasound-guided FNA. *Cancer* 2006; 108:180-185.

[11] Nguyen P, Feng JC, Chang KJ. Endoscopic ultrasound (EUS) and EUS-guided fine-needle aspiration (FNA) of liver lesions. *Gastrointestinal endoscopy* 1999; 50:357-361.

[12] Gimeno-Garcia AZ, Elwassief A. How to improve the success of endoscopic ultrasound guided fine needle aspiration cytology in the diagnosis of pancreatic lesions. *Journal of interventional gastroenterology*; 2:31-36.

[13] Suits J, Frazee R, Erickson RA. Endoscopic ultrasound and fine needle aspiration for the evaluation of pancreatic masses. *Arch. Surg.* 1999; 134:639-642; discussion 642-633.

[14] Wittmann J, Kocjan G, Sgouros SN, Deheragoda M, Pereira SP. Endoscopic ultrasound-guided tissue sampling by combined fine needle aspiration and trucut needle biopsy: a prospective study. *Cytopathology* 2006; 17:27-33.

[15] Ryozawa S, Kitoh H, Gondo T, et al. Usefulness of endoscopic ultrasound-guided fine-needle aspiration biopsy for the diagnosis of pancreatic cancer. *Journal of gastroenterology* 2005; 40:907-911.

[16] Voss M, Hammel P, Molas G, et al. Value of endoscopic ultrasound guided fine needle aspiration biopsy in the diagnosis of solid pancreatic masses. *Gut* 2000; 46:244-249.

[17] Bhutani MS, Hawes RH, Baron PL, et al. Endoscopic ultrasound guided fine needle aspiration of malignant pancreatic lesions. *Endoscopy* 1997; 29:854-858.

[18] Gress F, Gottlieb K, Sherman S, Lehman G. Endoscopic ultrasonography-guided fine-needle aspiration biopsy of suspected pancreatic cancer. *Annals of internal medicine* 2001; 134:459-464.

[19] Levy MJ. Know when to biopsy 'em, know when to walk away. *Gastrointestinal endoscopy* 2006; 63:630-634.

[20] Horwhat JD, Paulson EK, McGrath K, et al. A randomized comparison of EUS-guided FNA versus CT or US-guided FNA for the evaluation of pancreatic mass lesions. *Gastrointestinal endoscopy* 2006; 63:966-975.

[21] Micames C, Jowell PS, White R, et al. Lower frequency of peritoneal carcinomatosis in patients with pancreatic cancer diagnosed by EUS-guided FNA vs. percutaneous FNA. *Gastrointestinal endoscopy* 2003; 58:690-695.

[22] Levy MJ, Wiersema MJ. EUS-guided Trucut biopsy. *Gastrointestinal endoscopy* 2005; 62:417-426.

[23] Iglesias-Garcia J, Poley JW, Larghi A, et al. Feasibility and yield of a new EUS histology needle: results from a multicenter, pooled, cohort study. *Gastrointestinal endoscopy*; 73:1189-1196.
[24] AJCC Cancer Staging Manual. Edge S.B., Byrd D.R., Compton C.C., Fritz A.G., Greene F.L., Trotti A., Eds.; 7th ed. 2010: 241-246.
[25] Rosch T, Braig C, Gain T, et al. Staging of pancreatic and ampullary carcinoma by endoscopic ultrasonography. Comparison with conventional sonography, computed tomography, and angiography. *Gastroenterology* 1992; 102:188-199.
[26] Yasuda K, Mukai H, Fujimoto S, Nakajima M, Kawai K. The diagnosis of pancreatic cancer by endoscopic ultrasonography. *Gastrointestinal endoscopy* 1988; 34:1-8.
[27] Buchs NC, Chilcott M, Poletti PA, Buhler LH, Morel P. Vascular invasion in pancreatic cancer: Imaging modalities, preoperative diagnosis and surgical management. *World J. Gastroenterol.*; 16:818-831.
[28] Soriano A, Castells A, Ayuso C, et al. Preoperative staging and tumor resectability assessment of pancreatic cancer: prospective study comparing endoscopic ultrasonography, helical computed tomography, magnetic resonance imaging, and angiography. *The American journal of gastroenterology* 2004; 99:492-501.
[29] Dufour B, Zins M, Vilgrain V, et al. [Comparison between spiral x-ray computed tomography and endosonography in the diagnosis and staging of adenocarcinoma of the pancreas. Clinical preliminary study]. *Gastroenterologie clinique et biologique* 1997;21:124-130.
[30] Rivadeneira DE, Pochapin M, Grobmyer SR, et al. Comparison of linear array endoscopic ultrasound and helical computed tomography for the staging of periampullary malignancies. *Annals of surgical oncology* 2003; 10:890-897.
[31] Dewitt J, Devereaux BM, Lehman GA, Sherman S, Imperiale TF. Comparison of endoscopic ultrasound and computed tomography for the preoperative evaluation of pancreatic cancer: a systematic review. *Clin. Gastroenterol. Hepatol.* 2006; 4:717-725; quiz 664.
[32] Tio TL, Coene PP, den Hartog Jager FC, Tytgat GN. Preoperative TNM classification of esophageal carcinoma by endosonography. *Hepato-gastroenterology* 1990; 37:376-381.

[33] Bhutani MS, Hawes RH, Hoffman BJ. A comparison of the accuracy of echo features during endoscopic ultrasound (EUS) and EUS-guided fine-needle aspiration for diagnosis of malignant lymph node invasion. *Gastrointestinal endoscopy* 1997; 45:474-479.

[34] Akahoshi K, Chijiiwa Y, Nakano I, et al. Diagnosis and staging of pancreatic cancer by endoscopic ultrasound. *The British journal of radiology* 1998; 71:492-496.

[35] Legmann P, Vignaux O, Dousset B, et al. Pancreatic tumors: comparison of dual-phase helical CT and endoscopic sonography. *Ajr* 1998; 170:1315-1322.

[36] Hidalgo M. Pancreatic cancer. *The New England journal of medicine*; 362:1605-1617.

[37] Saftoiu A, Vilmann P. Role of endoscopic ultrasound in the diagnosis and staging of pancreatic cancer. *J. Clin. Ultrasound* 2009; 37:1-17.

[38] Singh M, Maitra A. Precursor lesions of pancreatic cancer: molecular pathology and clinical implications. *Pancreatology* 2007; 7:9-19.

[39] Sohn TA, Yeo CJ, Cameron JL, et al. Resected adenocarcinoma of the pancreas-616 patients: results, outcomes, and prognostic indicators. *J. Gastrointest. Surg.* 2000; 4:567-579.

[40] Michl P, Pauls S, Gress TM. Evidence-based diagnosis and staging of pancreatic cancer. *Best Pract. Res. Clin. Gastroenterol* 2006;20:227-251.

[41] Al-Haddad M, Eloubeidi MA. Interventional EUS for the diagnosis and treatment of locally advanced pancreatic cancer. Jop; 11:1-7.

[42] Jensen RT, Cadiot G, Brandi ML, et al. ENETS Consensus Guidelines for the management of patients with digestive neuroendocrine neoplasms: functional pancreatic endocrine tumor syndromes. *Neuroendocrinology*; 95:98-119.

[43] Balci NC, Semelka RC. Radiologic features of cystic, endocrine and other pancreatic neoplasms. *European journal of radiology* 2001; 38:113-119.

[44] Mansour JC, Chen H. Pancreatic endocrine tumors. *The Journal of surgical research* 2004; 120:139-161.

[45] Metz DC, Jensen RT. Gastrointestinal neuroendocrine tumors: pancreatic endocrine tumors. Gastroenterology 2008;135:1469-1492.

[46] Vanderveen K, Grant C. Insulinoma. *Cancer treatment and research*; 153:235-252.

[47] Pais SA, Al-Haddad M, Mohamadnejad M, et al. EUS for pancreatic neuroendocrine tumors: a single-center, 11-year experience. *Gastrointestinal endoscopy*;71:1185-1193.

[48] Faure JP, Tuech JJ, Richer JP, et al. Pancreatic metastasis of renal cell carcinoma: presentation, treatment and survival. *The Journal of urology* 2001; 165:20-22.

[49] Roland CF, van Heerden JA. Nonpancreatic primary tumors with metastasis to the pancreas. Surgery, gynecology and obstetrics 1989; 168:345-347.

[50] Z'Graggen K, Fernandez-del Castillo C, Rattner DW, Sigala H, Warshaw AL. Metastases to the pancreas and their surgical extirpation. *Arch. Surg.* 1998; 133:413-417; discussion 418-419.

[51] Klein KA, Stephens DH, Welch TJ. CT characteristics of metastatic disease of the pancreas. *Radiographics* 1998; 18:369-378.

[52] Yazbek T, Gayet B. The place of enucleation and enucleo-resection in the treatment of pancreatic metastasis of renal cell carcinoma. *Jop*; 13:433-438.

[53] Sweeney AD, Fisher WE, Wu MF, Hilsenbeck SG, Brunicardi FC. Value of pancreatic resection for cancer metastatic to the pancreas. *The Journal of surgical research*; 160:268-276.

[54] Palazzo L, Borotto E, Cellier C, et al. Endosonographic features of pancreatic metastases. *Gastrointestinal endoscopy* 1996; 44:433-436.

[55] DeWitt J, Jowell P, Leblanc J, et al. EUS-guided FNA of pancreatic metastases: a multicenter experience. *Gastrointestinal endoscopy* 2005; 61:689-696.

[56] Freeman C, Berg JW, Cutler SJ. Occurrence and prognosis of extranodal lymphomas. *Cancer* 1972; 29:252-260.

[57] Volmar KE, Routbort MJ, Jones CK, Xie HB. Primary pancreatic lymphoma evaluated by fine-needle aspiration: findings in 14 cases. *American journal of clinical pathology* 2004; 121:898-903.

[58] Behrns KE, Sarr MG, Strickler JG. Pancreatic lymphoma: is it a surgical disease? *Pancreas* 1994; 9:662-667.

[59] Rossi ED, Larghi A, Verna EC, et al. Endoscopic ultrasound-guided fine-needle aspiration with liquid-based cytologic preparation in the diagnosis of primary pancreatic lymphoma. *Pancreas*; 39:1299-1302.

[60] Naito Y, Okabe Y, Kawahara A, et al. Guide to diagnosing primary pancreatic lymphoma, B-cell type: immunocytochemistry improves the diagnostic accuracy of endoscopic ultrasonography-guided fine needle aspiration cytology. *Diagnostic cytopathology*; 40:732-736.

[61] Hernández Álvarez-Buylla N, Gimeno García AZ, Ortega Sánchez JA, Quintero E. Role of endoscopic ultrasonography with and without fine needle aspiration cytology in the diagnosis and staging of lymphoma *Journal of Hematological Malignancies* 2012; 2: 55-62.

[62] Nayer H, Weir EG, Sheth S, Ali SZ. Primary pancreatic lymphomas: a cytopathologic analysis of a rare malignancy. *Cancer* 2004; 102:315-321.

[63] Abraham SC, Klimstra DS, Wilentz RE, et al. Solid-pseudopapillary tumors of the pancreas are genetically distinct from pancreatic ductal adenocarcinomas and almost always harbor beta-catenin mutations. *The American journal of pathology* 2002; 160:1361-1369.

[64] Papavramidis T, Papavramidis S. Solid pseudopapillary tumors of the pancreas: review of 718 patients reported in English literature. *Journal of the American College of Surgeons* 2005; 200:965-972.

[65] Tang LH, Aydin H, Brennan MF, Klimstra DS. Clinically aggressive solid pseudopapillary tumors of the pancreas: a report of two cases with components of undifferentiated carcinoma and a comparative clinicopathologic analysis of 34 conventional cases. *The American journal of surgical pathology* 2005; 29:512-519.

[66] Martin RC, Klimstra DS, Brennan MF, Conlon KC. Solid-pseudopapillary tumor of the pancreas: a surgical enigma? *Annals of surgical oncology* 2002; 9:35-40.

[67] Jani N, Dewitt J, Eloubeidi M, et al. Endoscopic ultrasound-guided fine-needle aspiration for diagnosis of solid pseudopapillary tumors of the pancreas: a multicenter experience. *Endoscopy* 2008; 40:200-203.

[68] Loya AC, Ratnakar KS, Shastry RA. Combined osteoclastic giant cell and pleomorphic giant cell tumor of the pancreas: a rarity. An immunohistochemical analysis and review of the literature. *Jop* 2004; 5:220-224.

[69] Watanabe M, Miura H, Inoue H, et al. Mixed osteoclastic/pleomorphic-type giant cell tumor of the pancreas with ductal adenocarcinoma: histochemical and immunohistochemical study with review of the literature. *Pancreas* 1997; 15:201-208.

[70] Nai GA, Amico E, Gimenez VR, Guilmar M. Osteoclast-like giant cell tumor of the pancreas associated with mucus-secreting adenocarcinoma. Case report and discussion of the histogenesis. *Pancreatology* 2005; 5:279-284.

[71] Moore JC, Hilden K, Bentz JS, Pearson RK, Adler DG. Osteoclastic and pleomorphic giant cell tumors of the pancreas diagnosed via EUS-guided FNA: unique clinical, endoscopic, and pathologic findings in a series of 5 patients. *Gastrointestinal endoscopy* 2009; 69:162-166.

[72] -Yan BM, Pai RK, Van Dam J. Diagnosis of pancreatic gastrointestinal stromal tumor by EUS guided FNA. *Jop* 2008; 9:192-196.

[73] Yu RS, Sun JZ. Pancreatic schwannoma: CT findings. *Abdominal imaging* 2006; 31:103-105.

[74] Rao RN, Vij M, Singla N, Kumar A. Malignant pancreatic extra-gastrointestinal stromal tumor diagnosed by ultrasound guided fine needle aspiration cytology. A case report with a review of the literature. *Jop*; 12:283-286.

[75] Suzuki S, Kaji S, Koike N, et al. Pancreatic schwannoma: a case report and literature review with special reference to imaging features. *Jop*; 11:31-35.

[76] Pausawasdi N, Apisarnthanarak P, Pongpaibul A, Charatcharoenwitthaya P. Pancreatic lipoma diagnosed by EUS-FNA. *Gastrointestinal endoscopy*;76:668-669.

[77] Suzuki R, Irisawa A, Hikichi T, et al. Pancreatic lipoma diagnosed using EUS-FNA. A case report. *Jop* 2009; 10:200-203.

In: Ultrasonography
Editor: Alexander Burdenko

ISBN: 978-1-62417-536-7

Chapter 4

THE SUGGESTION FOR STANDARD SCANNING PROCEDURES OF ULTRASONOGRAPHIC EXAMINATIONS WHEN SURVEYING CERVICAL LYMPH NODES IN PATIENTS WITH ORAL SQUAMOUS CELL CARCINOMA

Nao Wakasugi-Sato[1], Masafumi Oda[1], Tatsurou Tanaka[1], Shinji Kito[1], Shinobu Matsumoto-Takeda[1], Shinya Kokuryo[2], Manabu Habu[2], Masaaki Kodama[2], Kensuke Yamauchi[2], Shinnosuke Nogami[2], Daigo Yoshiga[2], Ikuya Miyamoto[2], Izumi Yoshioka[3], Yoshihiro Yamashita[4], Noriaki Yamamoto[5], Tetsuro Wakasugi[6], Kou Matsuo[7], Ayataka Ishikawa[7], Yuji Seta[7], Kazuhiro Tominaga[2] and Yasuhiro Morimoto[1,5,]*

* Send all correspondence to: Yasuhiro Morimoto DDS PhD, Division of Diagnostic Radiology, Department of Oral Diagnostic Science, Kyushu Dental College, 2-6-1 Manazuru, Kokurakita-ku, Kitakyushu 803-8580, Japan. Tel: 81-93-285-3094, Fax: 81-93-285-3094. E-mail: rad-mori@kyu-dent.ac.jp.

[1]Department of Oral Diagnostic Science,
Kyushu Dental College, Kitakyushu, Japan
[2]Department of Oral and Maxillofacial Surgery,
Kyushu Dental College, Kitakyushu, Japan
[3]Department of Oral and Maxillofacial Surgery,
Miyazaki University, Miyazaki, Japan
[4]Department of Oral and Maxillofacial Surgery,
Fukuoka Dental College, Fukuoka, Japan
[5]Department of Oral and Maxillofacial Surgery,
Oita University, Oita, Japan
[6]Department of Otorhinology,
University of Occupational and Environmental Health,
Kitakyushu, Japan
[7]Department of Oral Bioscience,
Kyushu Dental College, Kitakyushu, Japan
[8]Center for Oral Biological Research,
Kyushu Dental College, Kitakyushu, Japan

ABSTRACT

In our previous study, we elucidated that scans that included the thyroid gland should be performed when surveying the cervical lymph nodes for metastasis during ultrasonographic (US) examinations in patients with oral squamous cell carcinoma (SCC). However, no standard procedures have yet been established for US examinations to survey the cervical lymph nodes in patients with oral SCC. Therefore, we developed a standard US procedure and evaluated the legitimacy and drawbacks associated with it. For our US scan procedure to survey cervical lymph nodes in patients with oral SCC, five axial, two cross-sectional, and one coronal tomographic scan using a probe with a major axis of 10 cm and a minor axis of 2 cm were used for the lateral necks of Japanese patients. If the resulting images were thought to be insufficient by the examiner, then some additional and repeated US examinations were performed.

In the present study, we analyzed the detection rates of cervical lymph nodes with a major axis over 8 mm using our US procedure for patients with oral SCC as a gold standard, using both magnetic resonance (MR) images and computed tomography (CT) images, which were obtained by both an expert and by a less-experienced examiner to examine how difficult the procedure was to follow and how accurate the results were. At the same time, the mean scanning hours spent by the expert and less-experienced examiner per patient was assessed when using our US procedure. In addition, the ink-smeared areas on the volunteer's neck as a result of our US procedure were analyzed. The detection rates of enlarged cervical lymph nodes using our US procedure were almost 95% for both the expert and the less-experienced examiner. In addition, the overlapped area of ink smeared on a standard volunteer's neck using our standard US procedure perfectly matched the lymph nodes in the cervical areas and thyroid glands. The time spent scanning the patients by the expert ranges from fifteen to twenty minutes, but the less-experienced examiner spent from twenty to twenty-five minutes per patient. Therefore, our US procedure for surveying the cervical lymph nodes, including thyroid glands, in patients with oral SCC can be recommended as a relatively good method, but inexperienced examiners should be prepared to spend additional time analyzing the scans.

Keywords: Ultrasonography; procedure; cervical lymph nodes; oral cancer; thyroid glands

INTRODUCTION

The identification of metastases in cervical lymph nodes has a major effect on the prognosis and treatment of patients with oral squamous cell carcinoma (SCC).[1, 2] Modalities such as computed tomography (CT) scans, magnetic resonance (MR) imaging, and ultrasonography (US) have been applied for the diagnosis of metastatic lymph nodes.[3-10] In particular, US, unlike CT scans and MR images, is a completely noninvasive and inexpensive technique that can be used to examine and follow up the changes in the cervical lymph nodes, and it is most often used for the short-term follow-up of patients.[1-10]

To our knowledge, however, there has been no publication describing standard procedures for US when surveying the cervical lymph nodes in patients with oral SCC. Therefore, the respective doctor or dentist generally develops their own method for such US scanning when surveying the cervical lymph nodes in patients with oral SCC. To provide more accurate and harmonized information about the diagnosis and prognosis of oral SCC, a standard US procedure is needed. In addition, the lack of standardization regarding the US methodology for cervical lymph nodes detection was due to the fact that we first examined only the particular one by US, rather than surveying the entire cervical area since US has generally been considered to be a supplementary technique in comparison to CT scans and MR images.

It is often difficult to evaluate the cervical lymph nodes by US in patients with oral SCC in general practice because there are major differences that have been reported based on the type of scanning procedure used. In a previous study, we elucidated the clinical significance of US examination for thyroid glands when surveying the metastasis of the cervical lymph nodes in patients with oral SCC [2], and next we should develop a standard scanning method for the US detection of cervical lymph nodes, including the thyroid glands, in patients with oral SCC.

In the present article, we introduce our US procedure for surveying the cervical lymph nodes for metastasis in patients with oral SCC, and report our evaluation of the legitimacy and problems associated with our procedure. To accomplish this, we evaluated the detection rates of metastatic cervical lymph nodes and the scanning times using our US procedures for patients with oral SCC. In addition, we identified the potential problems associated with our US procedure by comparing the detection rates and scanning times between an expert radiologist and an examiner who did not have very much experience.

Explanation of Our US Procedure for Surveying the Cervical Lymph Nodes for Metastasis in Patients with Oral SCC

Our US scan procedure for surveying the cervical lymph nodes of patients with oral SCC utilized a probe with a major axis of 10 cm and a minor axis of 2 cm (Figure 1), and basically obtained 5 axial, 2 cross-sectional, and 1 coronal tomographic scan of the lateral necks of Japanese patients. However, if it was concluded that there would be an insufficient scan area based on the examiner's decision, additional and repeated US examinations were undertaken. The details of our procedure are explained in the following sections, and are also shown in figures.

Patients were positioned on their back on medical beds. First, the proximal edge of the probe was placed at the center of the neck at the level of the sub-clavicle (Figure 2A). For the first scan, the probe should move up until the level of the divergence of the carotid arteries from the level of the sub-clavicle (Figure 2B). As the landmarks of the first scan, the carotid arteries at the end of the distal region of the probes can be continuously visualized (Figure 2C), and the thyroid glands at the lower levels were also included. In the first scan, lymph node metastases in the proximal areas of the internal jugular vein and various diseases of the thyroid glands should be identified. Next, the second scan should be performed to overlap with the distal half of the first scan, and it should continue up to the level of divergence of the carotid arteries from the level of the sub-clavicle. As the landmarks of the second scan, the carotid arteries at the center region of the probes can be continuously visualized (Figure 2D). The lymph nodes around the internal jugular vein should be diagnosed during the second scan.

The third scan should overlap the distal half of the second scan and should reach the level of divergence of the carotid arteries from the level of the sub-clavicle. As the landmarks of the third scan, the carotid arteries at the end of proximal region of the probes can be continuously visualized (Figure 2E). In the third scan, the lymph nodes in the distal areas of internal jugular vein should be carefully examined. The fourth scan should be performed to overlap the distal half of the third scan, and should continue up to the level of divergence of the carotid arteries from the level of the sub-clavicle. As the landmarks of the fourth scan, the sternocleidomastoid muscles could be continuously visualized from the distal position at the lower level to the proximal position at the upper level (Figure 2F). In the fourth scan, the lymph nodes in accessory nerves should be carefully examined.

The fifth scan should overlap the distal half of the fourth scan, and should continue to the level of divergence of the carotid arteries. As the landmarks of the fifth scan, the trapezius muscle can be continuously visualized from the proximal position at the lower level to the distal position at the upper level (Figure 2G). In the fifth scan, the lymph nodes around accessory nerves should be examined.

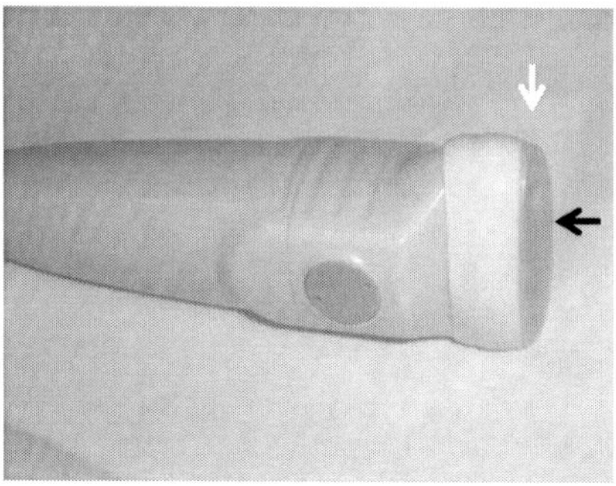

Figure 1. A photograph of a transducer of 7.5 MHz (Echo Camera SSD-1200CV; Aloka, Tokyo, Japan) that was used for the ultrasonographic assessments in the present study. The major axis (black arrows) of the probe was 10 cm, and the minor axis (white arrows) 2 cm.

For lymph nodes in the submental (Figure 3A) and submandibular (Figure 3B) regions, and within and around the parotid glands (Figure 3C), cross-sectional scans of the mandible should be continuously performed two times. As the landmarks of the sixth and seventh scans, the mylohyoid muscle (Figure 3D), submandibular glands (Figure 3E), and parotid glands (Figure 3F) can be visualized from the proximal to distal areas, in that order.

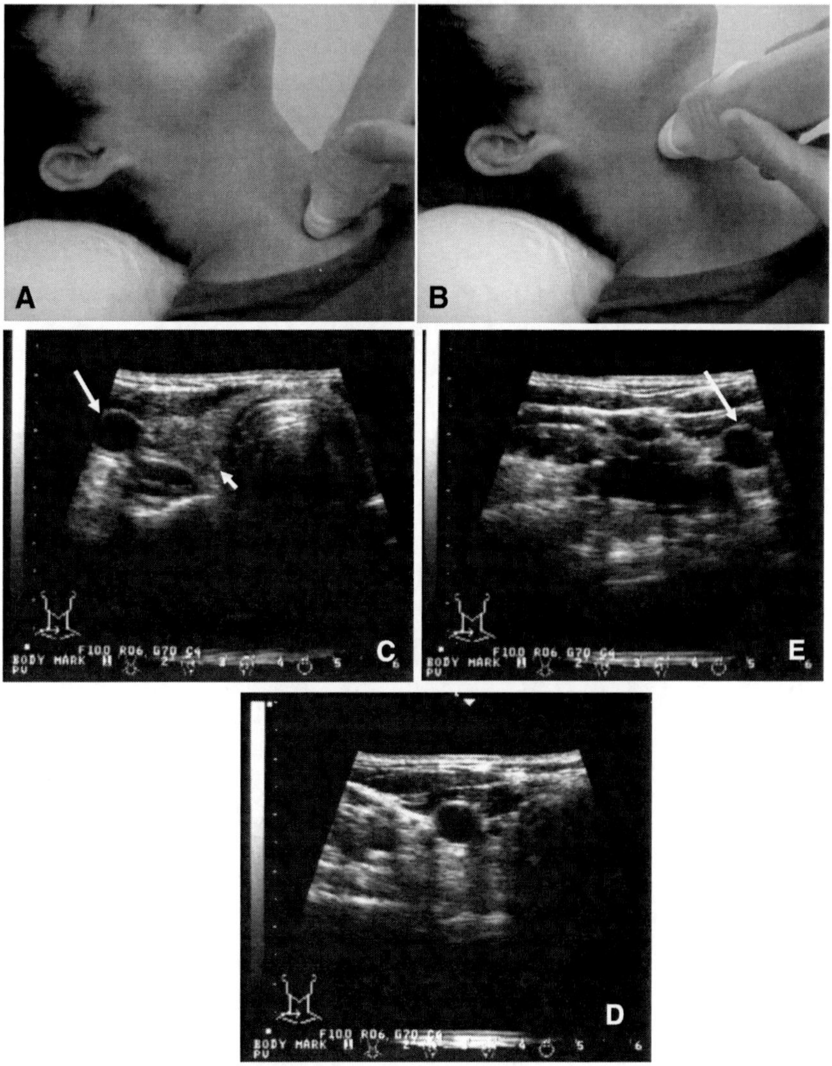

Figure 2. (Continued)

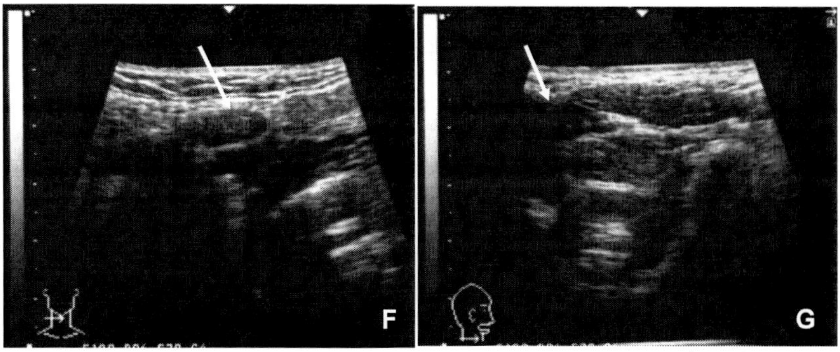

The patients were placed in a supine position on medical beds. First, the proximal edge of the probe was placed at the center of the neck at the level of the sub-clavicle (A). As the first scan, the probe should be scanned until the upper level of the divergence of the carotid arteries from the level of the sub-clavicle (B). As the landmarks of the first scan, the carotid arteries (arrow) at the end of the distal region of the probes can be continuously visualized (C), and the thyroid glands (arrowhead) at the lower levels are included. As the landmarks of the second scan, carotid arteries (arrow) at the center region of the probes can be continuously visualized (D). As the landmarks of the third scan, the carotid arteries (arrow) at the end of the proximal region of the probes can be continuously visualized (E). As the landmarks of the fourth scan, the sternocleidomastoid muscles (arrow) can be continuously visualized from the distal area at the lower level to the proximal area at the upper level (F). As the landmarks of the fifth scan, the trapezius muscle (arrow) can be continuously visualized from the proximal area at the lower level to the distal area at the upper level (G).

Figure 2. Photographs and US views obtained during the first to fifth scans using the standard US procedure for surveying the cervical lymph nodes for metastasis in patients with oral SCC.

In the sixth and seventh scans, the lymph nodes in the submental and submandibular areas, and within and around the parotid gland areas should be examined. The procedures mentioned above should be done for both the right and left sides of the neck.

Finally, coronal scans in the center of the submental regions should be performed (Figure 4A). As the landmarks of the eighth scan, the mylohyoid muscle, sublingual glands (Figure 4B), and geniohyoid muscle (Figure 4C) can be visualized from the frontal to posterior areas, in that order. In the eighth scan, the lymph nodes in the submental and sublingual areas should be examined. If there is thought to be an insufficient scan area based on the examiner's conclusions, then additional and/or repeated US examinations should be performed.

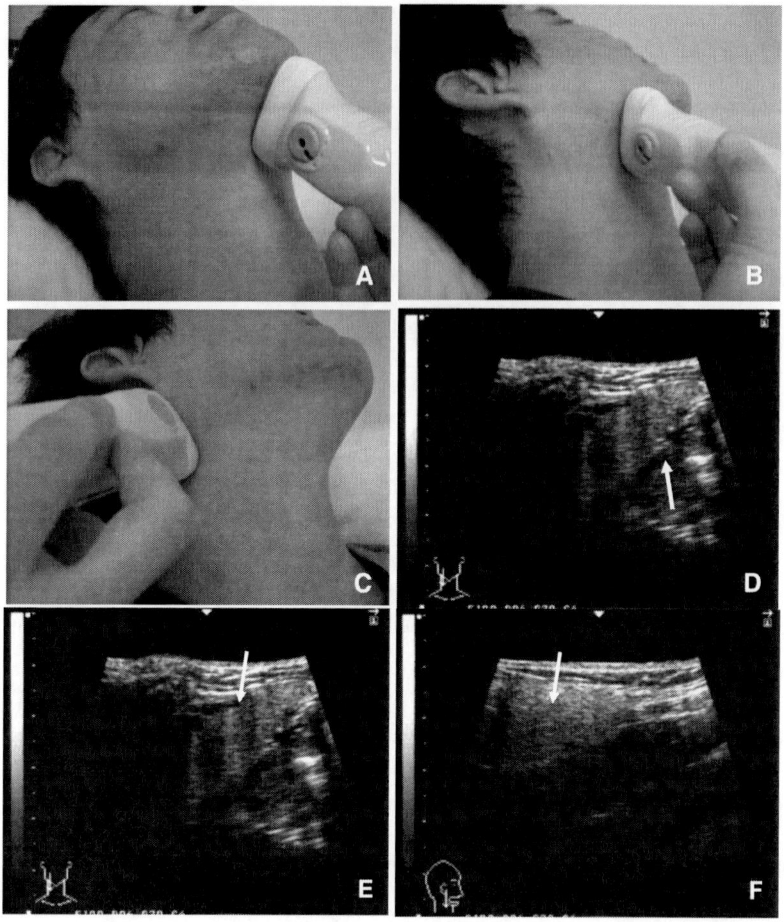

The cross-sectional scan for the mandible should be continuously done as the sixth and seventh scans for the lymph nodes in the submental area (A), submandibular area (B), and area within and around the parotid glands (C). As the landmarks of the sixth and seventh scans, the mylohyoid muscle (D: arrow), submandibular glands (E: arrow), and parotid glands (F: arrow) can be visualized from the proximal to distal areas, in that order.

Figure 3. Photographs and US views obtained during the sixth and seventh scans using the standard US procedure for surveying the cervical lymph nodes for metastasis in patients with oral SCC.

The detectable lymph nodes in the respective areas were drawn in mapping reports for each patient. The reports before mapping (Figure 5A) and after mapping (Figure 5B) are shown. In the mapping reports, the trachea, thyroid cartilage, thyroid glands, carotid arteries (including the internal and external carotid arteries), and mandibles were pre-described (Figure 5A). Detectable lymph nodes were added to the respective regions according to the results of the US scans (Figure 5B).

PATIENTS AND METHODS

The subjects consisted of 58 patients (31 males and 27 females) with SCC of the oral cavity, who underwent ultrasonographic examinations in all neck areas during their first examination for the detection of metastatic lymph nodes at Kyushu Dental College Hospital between 2011 and 2012. The Human Investigations Committee of Kyushu Dental College protected the individuals' rights for this study.

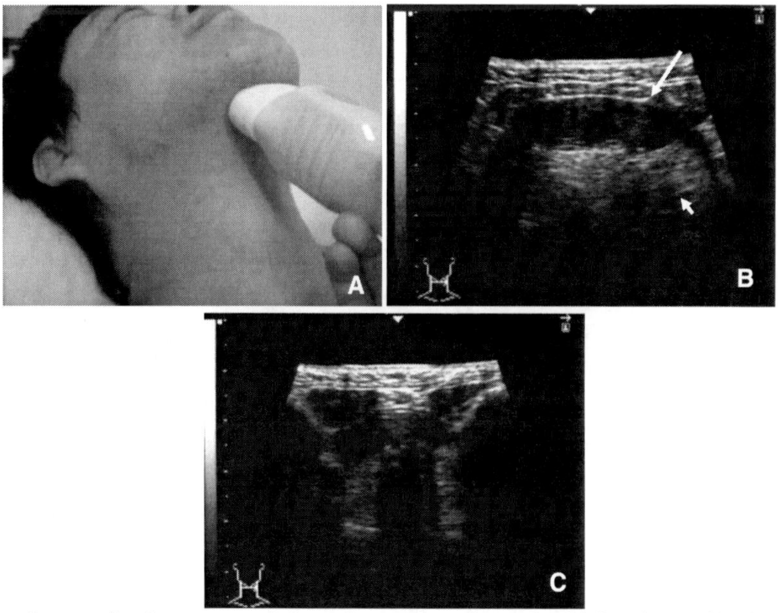

Coronal scans in the center of the submental regions should be done (A). As the landmarks of the eighth scan, the mylohyoid muscle (B: arrow), sublingual glands (B: arrowhead), and geniohyoid muscle (C: arrow), and can be visualized from the frontal to posterior areas, in that order.

Figure 4. Photographs and US views obtained during the eighth scan using the standard US procedure for surveying the cervical lymph nodes for metastasis in patients with oral SCC.

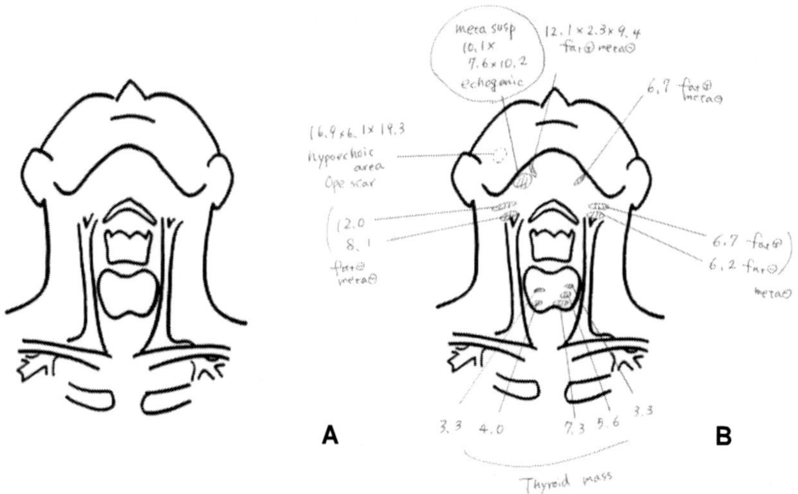

Figure 5. Photographs of the mapping reports for detectable lymph nodes in respective areas in each patient. In the mapping reports, the trachea, thyroid cartilage, thyroid glands, carotid arteries (including internal and external carotid arteries), and mandibles were pre-described (A). Detectable lymph nodes were added to the respective regions according to the results of the US scan (B).

A transducer of 7.5 MHz (Echo Camera SSD-1200CV; Aloka, Tokyo, Japan) was used for the ultrasonographic assessment. The major axis of the probe was 10 cm, and the minor axis was 2 cm, as mentioned above. The cervical lymph nodes detected according to our standard US procedures were compared to the detectable lymph nodes observed on MR images and CT scans, which were performed as the gold standard.

The CT scans were performed with the Aquilion™ device (Toshiba Co. Ltd., Tokyo, Japan) after the patient received an intravenous dose of 50 mL iohexol (300 mgI/mL; Omunipaque 300™, Daiichi Pharmaceutical Co. Ltd., Tokyo, Japan) at the start of scanning and an additional 50 mL intravenous infusion during the scan to allow for better visualization of the vascular structures. Scanning was performed in the axial plane without angulation, in 5 mm-thick contiguous sections from the cavernous sinuses to the thoracic inlet. Images were obtained with standard algorithms and soft-tissue windows. MR images were acquired using a 1.5 T full-body magnetic resonance system (EXCELART Vantage™ Powered by Atlas; Toshiba, Tokyo, Japan) equipped with a circularly polarized neck coil. We acquired both T1-weighted and T2-weighted images (axial and coronal sections). The imaging parameters used in these sequences are shown in Table 1.

Table 1. MR imaging parameters

	Sequences	
	T1WI	T2WI
TR (ms)	500	3500
TE (ms)	15	108
Flip angle	90	90
FOV (mm)	200 X 200	200 X 200
Section thickness (mm)	5.0	5.0
Slab thickness (mm)	100	100
Matrix (pixels)	224 X 320	224 X 256

Table 2. Detection rates in the cervical lymph nodes of each area by the expert or the less-experienced examiner during an ultrasonographic examination

Region	Expert	Less-experienced examiner
Internal jugular lymph nodes	97%	96%
Superior level	98%	98%
Middle level	85%	91%
Inferior level	100%	83%
Spinal accessory lymph nodes	100%	100%
Submental	93%	100%
Submandibular	93%	89%
Parotid glands	100%	100%
Total	95.9%	94.4%

Table 3. Differences in the scanning time between expert and less-experienced radiologists for detecting cervical lymph nodes using ultrasonographic examinations

	Radiologists	
	Expert	Less-experienced examiner
Total hours	16.0±1.8 min	21.4±5.8 min

min: Minute.

Based on the locations, sizes, figures, and internal characteristics of the lymph nodes, the lymph nodes detected by US were compared to those on CT and MRI. The detection rates of lymph node metastases by US were compared when they were analyzed by one expert radiologist (Y.M.) who had more than

17 years of experience and by another radiologist (N.W.) who had 5 years of experience using random sampling in the US examinations. Both radiologists were blinded to the CT scan and MR image findings. At the same time, the time spent analyzing the scans was also measured for each examiner. We classified the superior, middle, and inferior internal jugular lymph nodes, respectively, spinal accessory lymph nodes, as well as lymph nodes in the submental and submandibular areas, and around the parotid glands areas. The detection rates were compared for each area. We selected lymph nodes with a major axis of 8 mm as the detectable lymph nodes, because it was difficult for us to see the correlations between the lymph nodes with US and lymph nodes using CT scan and MR images. In order to evaluate the impact of experience with US examinations on the detection of cervical lymph nodes, we compared the detection rates and scanning time per patient between the expert and the less experienced radiologist.

Next, to confirm the appropriateness of our US procedure, we evaluated the degree of neck covered with coloring matter by our US scan for surveying the cervical lymph nodes for metastasis during US examinations in patients with oral SCC using five Japanese volunteers with an average body size. The measurements around the neck in these five volunteers ranged from 40.0 to 45.3 cm. Informed consent was obtained from all five volunteers before examinations.

All statistical analyses were performed using SPSS version 11 statistical software (SPSS Inc., Chicago, IL, USA). Differences between the two groups were analyzed by student t-test. Categorical variables were compared by the χ^2 test or Fisher exact test.

Results

The Detection Rate and Scanning Time for Ultrasonographic Examinations Surveying the Cervical Lymph Nodes of Patients with Oral SCC

The detection rates of detectable lymph nodes with a major axis of 8 mm or larger as determined using US with our standard scanning procedure are shown in Tables 2 and 3. The total detection rate by an expert radiologist using US was 95.9±5.5% (mean±SD) for ten patients with oral SCC, and the mean scanning time was 16.0±1.8 minutes (mean±SD) per patient.

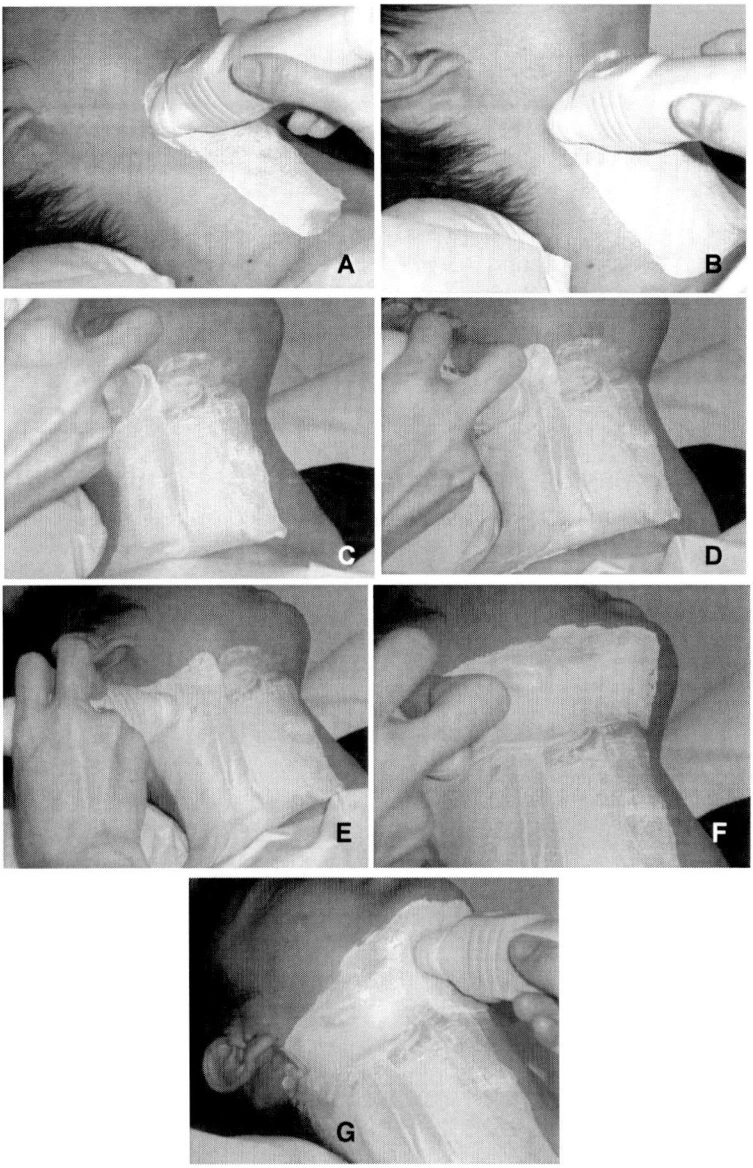

Neck areas smeared by the first US scan (A), the second (B), the third (C), the forth (D), the fifth (E), the sixth (F), the seventh (F), the eighth (G) were visualized. The area of overlap between the ink smeared during our standard US scan perfectly matched the lymph nodes in the cervical areas and thyroid glands (G).

Figure 6. Photographs of overlapping area of the ink smeared on the volunteer's neck by our standard US scan with the lymph nodes in all five volunteers (A-G).

There were no significant differences among the different areas. The total detection rate by the less-experienced examiner was 94.4±6.8% (mean±SD) for ten patients with oral SCC. The scanning time was 21.4±5.8 minutes (mean±SD) per patient, and there was no significant differences among the different areas that were scanned.

The detection rate of the expert was similar to that of the less-experienced examiner (Table 2). No significant difference in the detection rates in the respective areas was found between the expert and the less-experienced examiner (Table 2). However, the scanning time for the expert was significantly shorter than that of the less-experienced examiner ($P<0.00001$; Table 3).

Areas of Overlap for the Ink Smeared on the Volunteer's Neck during the Standard US Scan and the Locations of Lymph Nodes

The area of overlap between the ink smeared during our standard US scan perfectly matched the lymph nodes in the cervical areas and thyroid glands in all five volunteers (Figure 6).

DISCUSSION

Many publications have demonstrated that US is a useful modality for the examination and follow up of metastatic cervical lymph nodes in patients with oral SCC.[11-14] To our knowledge, however, there were no previously published standard procedures for the detection of cervical lymph nodes by US. In the present study, we described and advocated the use of our standard US scanning method for surveying the cervical lymph nodes in patients with oral SCC, and also reported the utility, appropriateness and ease of using our US method.

As the detection ratios in cervical lymph nodes were almost 100% in all areas when our method was used, we speculated that our procedure may be sufficient as the standard for surveying cervical lymph nodes. If the performance of axial, coronal, and cross sectional scanning per neck would increase, then only the length of examination time should become longer. If it was thought that there would be an insufficient scan area for a patient based on an examiner's decision, additional and/or repeated US examinations were then performed.

So far, we have had no misdiagnoses due to overlooking metastatic cervical lymph nodes using our method. In fact, the detection rates were about 95% in the present study. A possible explanation for these good results may be that our detectable cervical lymph nodes on US were different from those on CT and MRI because of the presence of a gap in the tomographic planes between US and CT or MRI.

Unfortunately, we could not definitively determine whether or not our US procedure and times were appropriate, because we were unable to directly compare our method with another method. Our results only reflect the standard for an US survey of cervical lymph nodes in patients with oral SCC, but may be useful for patients with other cervical lymph node diseases, including malignant lymphoma, multiple lymphadenitis, and so on. We expect that many similar reports from various hospitals will be published, and that the standardization of a US method for the cervical lymph nodes will thus eventually be established based on these many reports and future prospective clinical trials.

Using our US scanning method, the detection rate of the less-experienced radiologist was almost equal to that of the experienced examiner. In other words, there were no differences in the diagnosis of the cervical lymph nodes between the expert and less-experienced radiologists. In addition, no significant difference of the detection rates in the various areas was found between the expert and the less-experienced examiner. Our present results may help in the development of a standard method for surveying the cervical lymph nodes in patients with oral SCC using US. Based on the present results, we advocate that our US procedure should be selected as the standard method for surveying the cervical lymph nodes of patients with oral SCC. In the present study, the less-experienced examiner had performed US scans for 5 years, and had experienced more than 300 cases where the cervical lymph nodes of patients with oral SCC were scanned. If beginner radiologists (with less experience) had been included in the study, there might have been a significant difference in the detection rate of cervical lymph nodes between the experts and beginners. In particular, based on our empirical knowledge, the difference in the detection rates in the internal jugular, spinal accessory, and submandibular lymph nodes would likely have been lower. Such areas may tend to be blind spots because they are difficult to evaluate because of the lack of contact between the probe and skin and due to their being farther from scanning centers. Therefore, at our institution, we have instructed beginners that there is a tendency to overlook lateral internal jugular, spinal accessory, and submandibular lymph nodes when surveying cervical lymph nodes by US in patients with oral SCC.

Despite the minimal differences between the expert and less-experienced radiologists, the length of the analysis of the scans for the expert was significantly shorter than that of the less-experienced examiner, namely by about 5-10 minutes per patient. It is possible that, for the examiner who had less experience, there were likely to be additional and repeated scans, due to unskilled handling methods, and the time required to analyze the scans may be longer because of the lack of experience. Of course, the longer scanning time for beginners would improve by appropriate instruction by an experienced radiologist.

In previous reports, we elucidated that patients with oral SCC have a relatively high rate of abnormal ultrasonographic findings in the thyroid glands, and scans including the thyroid glands should be taken when surveying cervical lymph nodes by US for the detection of metastasis.[2, 6] It is also necessary to predict other general health conditions based on the oral-related examinations as much as possible to avoid the need for more invasive diagnostic methods. Therefore, we should scan all lymph nodes in the cervical areas, including the bilateral thyroid glands. As a result, the patients can be more confident in their diagnosis, and will therefore have greater faith in radiologists. In addition, a variety of medical problems will likely be able to be diagnosed, along with the need for more diverse medications, thus leading to a decrease in subsequent medical expenses due to the earlier and easier diagnosis. This is especially important as the number of aged persons is increasing in many nations, including Japan.

A potential disadvantage of our method is that we simplified the US scanning methods in many scanning directions (axial, sagittal, coronal, oblique, and so on).[12] As mentioned above, to our knowledge, there had been no publication about standard procedures for US when surveying the cervical lymph nodes in patients with oral SCC. Therefore, we selected the axial mode as the scanning method for the internal jugular and spinal accessory lymph nodes based on the standard method of US scanning using a transducer with a major axis of 10 cm. In fact, the axial US scanning from the upper parts of the clavicle to the inferior part of the mandible is currently performed for surveying the internal jugular and spinal accessory lymph nodes in patients with oral SCC. For the submental, submandibular, and parotid lymph nodes, the transducers were set to scan parallel to the mandibles. In addition, we recommend performing coronal scans in the center of the submental regions. Therefore, whenever we detect lymph nodes with a major axis of 7 or 8 mm, we further examine the nodes, noting the sizes, figures, and internal characteristics of the lymph nodes, and add them into the mapping reports.

Using our US procedure, the detection rates of affected cervical lymph nodes by US was almost 100%, which would be useful for routine clinical assessments. In addition, the scanning time was about 20 minutes for surveying all cervical lymph nodes by US. Therefore, our scanning procedure appears to be appropriate for the detection of cervical lymph nodes by US in patients with oral SCC. Of course, the precise scanning procedure and time will vary based on the state of each individual patient, including their general physical condition, physique, number of lymph nodes, presence or absence of metastasis, and so on.

In the present study, we used the flat probe with a 10 cm maximum axis, which is the most popular cervical probe used in Japan. If other probes with longer and more appropriate sizes, based on the patient's individual features, were selected, the scanning times would be shorter. In addition, we speculate that the scanning times are likely dependent on the patients' body size and type. Therefore, we think that investigation on the size and type of the probe and different patient body types should also be performed in future studies to provide further standardization of these procedures.

Conflict of Interest Statement

No potential conflicts of interest were disclosed.

Sources of Support in the Form of Grants

This study was supported in part by grants-in-aid for scientific research from the Ministry of Education, Science, Sports and Culture of Japan and from Kitakyushu City to YM.

References

[1] Snow GB, Annyas AA, van Slooten EA, Bartelink H, Hart AA. Prognostic factors of neck node metastasis. *Clin. Otolaryngol. Allied Sci.* 1982; 7: 185-92.

[2] Wakasugi-Sato N, Wakasugi T, Oda M, Yamashita Y, Yoshioka I, Yamamoto N, Habu M, Kodama M, Kokuryo S, Ichimiya H, Miyamoto I, Tanaka T, Kito S, Matsumoto-Takeda S, Ishikawa A, Seta Y, Matsuo K, Takahashi T, Tominaga K, Morimoto Y. Clinical significance of ultrasonographic examination including detection of thyroid gland diseases when surveying cervical lymph nodes in subjects with oral squamous cell carcinoma. *Oral Surg. Oral Med. Oral Pathol. Oral Radiol. Endod.* 2010; 109: e78-85.

[3] Morimoto Y, Tanaka T, Yoshioka I, Yamashita Y, Hirashima S, Kodama M, Ariyoshi W, Tomoyose T, Furuta N, Habu M, Okabe S, Kito S, Oda M, Kuroiwa H, Wakasugi-Sato N, Takahashi T, Tominaga K. New diagnostic imaging modalities for oral cancers (Chapter 5). In Oral Cancer Research Advances. 125-153, Nova Science Publishers, 2007.

[4] Morimoto Y, Tanaka T, Yamamoto N, Kodama M, Seta Y, Habu M, Oda M, Kito S, Wakasugi-Sato N, Matsumoto-Takeda S, Fukai Y, Tokitsu T, Tomikawa M, Matoba K, Yamashita Y, Yoshioka I, Takahashi T, Tominaga K. New Trends and Advances in Oral and Maxillofacial Imaging. *Curr. Med. Imaging Rev*. 2009; 5: 226-237.

[5] Oda M, Tanaka T, Kito S, Seta Y, Ishikawa A, Kodama M, Yamamoto N, Habu M, Miyamoto I, Kokuryo S, Khanal A, Basugi A, Wakasugi-Sato N, Matsumoto-Takeda S, Hamai M, Mandai S, Takino S, Matsuo K, Yamashita Y, Yoshioka I, Sakoda S, Takahashi T, Tominaga K, Koga H, Morimoto Y. Recent advances of the diagnostic images for oral cancers. In Oral Cancer: Causes, Diagnosis and Treatment, Nova Science Publishers, 2011.

[6] Wakasugi-Sato N, Kodama M, Matsuo K, Yamamoto N, Oda M, Ishikawa A, Tanaka T, Seta Y, Habu M, Kokuryo S, Ichimiya H, Miyamoto I, Kito S, Matsumoto-Takeda S, Wakasugi T, Yamashita Y, Yoshioka I, Takahashi T, Tominaga K, Morimoto Y. Advanced clinical usefulness of ultrasonography for diseases in oral and maxillofacial regions. *Int. J. Dent*. 2010: 639382.

[7] Ward-Booth RP, Williams ED, Faulkner TP, Earl PD. Ultrasound: a simple noninvasive examination of cervical swellings. *Plast. Reconstr. Surg*. 1984; 73: 577-81.

[8] de Bondt RB, Nelemans PJ, Hofman PA, Casselman JW, Kremer B, van Engelshoven JM, Beets-Tan RG. Detection of lymph node metastases in head and neck cancer: a meta-analysis comparing US, USgFNAC, CT and MR imaging. *Eur. J. Radiol*. 2007; 64: 266-72.

[9] Jank S, Robatscher P, Emshoff R, Strobl H, Gojer G, Norer B. The diagnostic value of ultrasonography to detect occult lymph node involvement at different levels in patients with squamous cell carcinoma in the maxillofacial region. *Int. J. Oral Maxillofac. Surg*. 2003; 32: 39-42.

[10] Esen G. Ultrasound of superficial lymph nodes. *Eur. J. Radiol*. 2006; 58: 345-59.

[11] Kagawa T, Yuasa K, Fukunari F, Shiraishi T, Miwa K. Quantitative evaluation of vascularity within cervical lymph nodes using Doppler ultrasound in patients with oral cancer: relation to lymph node size. *Dentomaxillofac. Radiol*. 2011; 40: 415-21.

[12] Differential diagnosis in head and neck imaging. Edited by Vogl TJ. Thieme. Stuttgart, 1999.

[13] Ariji Y, Kimura Y, Hayashi N, Onitsuka T, Yonetsu K, Hayashi K, Ariji E, Kobayashi T, Nakamura T. Power Doppler sonography of cervical lymph nodes in patients with head and neck cancer. *Am. J. Neuroradiol*. 1998; 19: 303-307.

[14] Yuasa K, Kawazu T, Kunitake N, Uehara S, Omagari J, Yoshiura K, Nakayama E, Kanda S. Sonography for the detection of cervical lymph node metastases among patients with tongue cancer: criteria for early detection and assessment of follow-up examination intervals. *Am. J. Neuroradiol*. 2000; 21: 1127-32.

In: Ultrasonography
Editor: Alexander Burdenko
ISBN: 978-1-62417-536-7

Chapter 5

PRE-ANALYTICAL ERROR: A FORGOTTEN RISK FOR USING ULTRASONOGRAPHY

Viroj Wiwanitkit
Wiwanitkit House, Bangkhae, Bangkok, Thailand
Hainan Medical University, China
Joseph Ayobabalola University, Nigeria

ABSTRACT

Ultrasonography is widely used as diagnostic technology in medicine at present. It is no doubt that ultrasonography has a lot of benefits, however, the risk in using this technique has to be mentioned. Based on the laboratory medicine principle, the concern on the pre-analytical phase for using diagnostic ultrasonography should be mentioned. The error in pre-analytical phase can be expected and this can be risk. Of interest, pre-analytical error in ultrasonography is usually not mentioned and it is an actual forgotten issue. This short commentary will summarize and discuss on this specific topic.

INTRODUCTION

Ultrasonography is the applied usage of sound in imaging construction. This can be very useful in many works including to medicine. In medicine, ultrasonography is widely used as diagnostic technology. It is a main diagnostic tool in medicine at present. Due to the noninvasive nature and free from radiation, ultrasonography is considered safe. It is also easy – to – use. Hence, ultrasonography is implemented in all levels of medical care centers.

It is no doubt that ultrasonography has a lot of benefits, however, the risk in using this technique has to be mentioned. As a rule, no technology poses no risk. For ultrasonography, the consideration on general quality cycle in diagnostic medicine should not be forgotten. Based on the laboratory medicine principle, the concern on the pre-analytical phase for using diagnostic ultrasonography should be mentioned. The error in pre-analytical phase can be expected and this can be risk. Of interest, pre-analytical error in ultrasonography is usually not mentioned and it is an actual forgotten issue. This short commentary will summarize and discuss on this specific topic.

What is Pre-Analytical Phase? And What is Its Importance in Diagnostic Radiology

Basically, the concept of quality cycle is used in laboratory medicine. However, it can be and should be extensively used in all fileds of investigative or diagnostic medicine. Hence, the application in diagnostic radiology, which is an actual diagnostic procedure, is warranted. Focusing on quality cycle, there are three main phases: pre-analytical, analytical and post-analytical phase. The first phase, pre-analytical phase, means any activities before the actual diagnostic investigative procedure. The second phase, analytical phase is the actual diagnostic investigative procedure. The third phase, post-analytica phase, covers any acitivities after diagnostic investigative procedure. With the concept, the practitioner should not overlook the importance of each phase.

The consideration is on the error, which can occur at any phase. Due to the referencing report, the error in any phases of diagnosis can be seen and the highest rate of error occurs in pre-analytical phase [1 – 3]. It should be noted that the pre-analytical phase is the first phase prior to the other phases. If there is any problem in this phase, the correctness and reliability of the investigative result cannot be expected. This is also the fact for the diagnostic radiology.

In diagnostic radiology, the pre-analytical phase is also existed. The pre-analytical phase for diagnostic ultrasonography includes many activities such as diagnostic ultrasonography requestration, patient preparation and delivery of the patient to the ultrasonography unit. For sure, the error or problem in the mentioned actitivies can be seen and the problem cannot be avoided if error occurs (Table 1).

Table 1. Some examples of errors in pre-analytical phase of diagnostic ultrasonography

Activities	Examples
1. Diagnostic ultrasonography requestration	Error in request form (incorrect examined site, incorrect technique, etc.) Misidentication of case
2. Patient preparation	Inappropriate preparation Misidentication of case
3. Delivery of the patient	Accident during delivery Misidentication of case

REPORTS ON PRE-ANALYTICAL PHASE ERROR OF DIAGNOSTIC ULTRASONOGRAPHY

Indeed, the error in diagnostic ultrasonography is not a new thing. There are many previous reports on this topic. However, it seems to be the forgotten issue. Most considerations on investigation quality control are usually on the diagnostic tool and diagnostic procedure. Fewer concerns are given to pre- and post- analytical phase. Here are some interesting reports on pre-analytical phae error of diagnostic ultrasonography (Table 2).

Table 2. Reports on pre-analytical phase error of diagnostic ultrasonography

Authors	Details
Eze et al. [4]	Eze et al. studied the errors in diagnostic ultrasonography that related to trainees. According to this audit, Eze et al. found that "Majority of the errors originate from lack of accurate filling of the patients request cards by the requesting physicians, lack of adequate preparation for the intended study, and typographical errors in writing reports [4]."
Zammit et al. [5]	Zammit et al. srudied the pitfalls in endorectal ultrasound for diagnosis of rectal tumor [5]. Zammit et al. reported that poor bowel preparation was an important cause of difficulty in performing endorectal ultrasound [5]. The inaccuracy of diagnostic ultrasonography due to the pitfall was also reported in this study [5].

Based on the described reports [4 – 5], it is no doubt that errors can be seen in pre-analytical phase of diagnostic ultrasonography. However, there is no specific systematic evaluation on the exact prevalence of this kind of error in diagnostic ultrasonography. This might reflect that this is an actual forgotten issue to be concerned.

RISK ON PRE-ANALYTICAL ERROR IN DIAGNOSTIC ULTRASONOGRAPJHY

If there is no problem from error, it is no use to deal with the error. As noted, the error in the pre-analytical phase of diagnostic ultrasonography is usually due to medical personnel, not the ultrasound analyzer. This might be different from analytical phase error of diagnostic ultrasonography, which depends mainly on the quality assessment program [6] and manufacturer support [7]. The potential of error to be occured in pre-analytical phase of diagnostic ultrasonography is confirmed. The error can result in problem and this can also result in risk. One might have a question on the risk that comes from pre-analytical error. The risk can be due to a) the incorrect diagnosis and b) the delayed diagnosis. The incorrect diagnosis might present as false posivity or false negativity. Either false negative or false positive result is not wanted. The false results due to the pre-analytical error can also be controlled similar to those one due to the analytical – error (which is usually due to the quality of the diagnostic ultrasonography tool). The incorrect diagnosis might lead to a) no proper therapeutic management to the patient or b) excessive and unnecessary therapeutic management to the patient. In some cases, the delayed diagnoses can also be observed and this can result in loss of curative chance of the patients.

Hence, the risk of pre-analytical error in diagnostic ultrasonography is a big issue to be considered. How to manage the risk is the new consideration. Quality control of the diagnostic ultrasonography must cover with risk management [8 – 9]. Basically, the risk management can be done via two main activities: preventive action and corrective action. The preventive action is any procedure for preventing the risk. This means pre-risk management. This is a good concept. All medical centers have to prepare specific protocols for prevention of the risk. Training of the practitioners who involve in the pre-analytical phase is needed. There should also be the rechecking system to support the routine work. The implementation of digital computational system might be helpful. Di Nallo et al. noted that "modern digital equipments used in serial measurements may be stable over the time and that calibrated phantom may enable preventive identification of corrective action [10]". Focusing on corrective action, it is the attempt to resolve the already existed problem. It might be a tertiary prevention. However, it is better than no action. The prompt management of the risk is required. Nevertheless, the thing that should not be forgotten is to record and report all problems. Analysis for the root cause of episode after corrective action for planning for preventive action for the future is recommended.

CONCLUSION

As a diagnostic tool, the concern on the quality and risk management in all phases for diagnostic ultrasonography is required. Focusing on the investigation cycle, the pre-analytical phase should be mentioned since it is the first phase with a high error. Since most of pre-analytical error is usually due to human error, good clinical practice, guideline and protocol as well as risk management system are all required for reducing the pre-analytical error in diagnostic ultrasonography.

REFERENCES

[1] Wiwanitkit V. Types and frequency of preanalytical mistakes in the first Thai ISO 9002:1994 certified clinical laboratory, a 6 - month monitoring. *BMC. Clin. Pathol.* 2001;1(1):5.

[2] Hawkins R. Managing the pre- and post-analytical phases of the total testing process. *Ann. Lab. Med.* 2012 Jan;32(1):5-16.

[3] Lippi G, Chance JJ, Church S, Dazzi P, Fontana R, Giavarina D, Grankvist K, Huisman W, Kouri T, Palicka V, Plebani M, Puro V, Salvagno GL, Sandberg S, Sikaris K, Watson I, Stankovic AK, Simundic AM. Preanalytical quality improvement: from dream to reality. *Clin. Chem. Lab. Med.* 2011 Jul;49(7):1113-26.

[4] Eze KC, Marchie TT, Eze CU. An audit of ultrasonography performed and reported by trainee radiologists. *West. Afr. J. Med.* 2009 Jul-Aug;28(4):257-61.

[5] Zammit M, Jenkins JT, Urie A, O'Dwyer PJ, Molloy RG. A technically difficult endorectal ultrasound is more likely to be inaccurate. *Colorectal. Dis.* 2005 Sep;7(5):486-91.

[6] Oh JK. Is core laboratory essential for using echocardiography in clinical trials? Controlled vs random error. *Eur. J. Echocardiogr.* 2002 Dec;3(4):245-7.

[7] Schwing C. Diagnostic error using ultrasound. Potential risks caused by deficient quality assurance by the manufacturer. *Fortschr. Med.* 1985 Apr 11;103(14):82-3.

[8] Insana MF, Hall TJ. Quality management of ultrasound diagnosis. *Clin. Diagn. Ultrasound.* 1993;28:161-81.

[9] Merz E, Bätzel A. Quality assurance in ultrasound diagnostics--where are we at the end of 2009? *Ultraschall. Med.* 2009 Dec;30(6):525-7.

[10] Di Nallo AM, Strigari L, Benassi M. A possible quality control protocol for Doppler ultrasound for organizational time optimization. *J. Exp. Clin. Cancer. Res.* 2006 Sep;25(3):373-81.

INDEX

C

D

E

F

G

H

N

O

P

Q

R

S

T

U

V

W

Y